The Social Life of Greylag Geese
Patterns, Mechanisms and Evolutionary Function in an
Avian Model System

The flock of greylag geese established by Konrad Lorenz in Austria
in 1973 has become an influential model animal system and one of
the few worldwide with complete life history data spanning several
decades. Based on the unique records of almost 1000 free-living greylag
geese, this is a synthesis of more than 20 years of behavioural research.
It provides a comprehensive overview of a complex bird society,
placing it in an evolutionary framework and drawing on a range
of approaches, including behavioural (personality, aggression, pair
bonding and clan formation), physiological, cognitive and genetic.

 With contributions from leading researchers, the chapters
provide valuable insights into historical and recent research
on the social behaviour of geese. All aspects of goose and bird
sociality are discussed in the context of parallels with mammalian
social organisation, making this a fascinating resource for anyone
interested in integrative approaches to vertebrate social systems.

ISABELLA B. R. SCHEIBER is a Researcher at the Behavioural
Ecology and Self-Organization Group of the University of Groningen,
The Netherlands, and an Associate Scientist at the Konrad Lorenz
Research Station (KLF), Grünau, Austria.

BRIGITTE M. WEIß is a visiting Researcher at the Institute of
Evolution and Ecology, University of Tübingen, Germany, and an
Associate Scientist at the KLF, Grünau, Austria.

JOSEF HEMETSBERGER is a Scientist at the Department of
Behavioural Biology, University of Vienna, and the KLF, Grünau,
Austria.

KURT KOTRSCHAL is Professor of Zoology at the Department of
Behavioural Biology, University of Vienna, and Director of the KLF,
Grünau, Austria.

The Social Life of Greylag Geese

Patterns, Mechanisms and Evolutionary Function in an Avian Model System

Edited by

ISABELLA B. R. SCHEIBER
University of Groningen, The Netherlands, and Konrad Lorenz Research Station, Grünau, Austria

BRIGITTE M. WEIß
University of Tübingen, Germany, and Konrad Lorenz Research Station, Grünau, Austria

JOSEF HEMETSBERGER
University of Vienna and Konrad Lorenz Research Station, Grünau, Austria

KURT KOTRSCHAL
University of Vienna and Konrad Lorenz Research Station, Grünau, Austria

CAMBRIDGE
UNIVERSITY PRESS

University Printing House, Cambridge CB2 8BS, United Kingdom

One Liberty Plaza, 20th Floor, New York, NY 10006, USA

477 Williamstown Road, Port Melbourne, VIC 3207, Australia

314-321, 3rd Floor, Plot 3, Splendor Forum, Jasola District Centre, New Delhi - 110025, India

79 Anson Road, #06-04/06, Singapore 079906

Cambridge University Press is part of the University of Cambridge.

It furthers the University's mission by disseminating knowledge in the pursuit of education, learning and research at the highest international levels of excellence.

www.cambridge.org
Information on this title: www.cambridge.org/9781108810555

© Cambridge University Press 2013

First published 2013
First paperback edition 2019

A catalogue record for this publication is available from the British Library

Library of Congress Cataloging in Publication data
The social life of greylag geese : patterns, mechanisms and evolutionary function in an avian model system / edited by Isabella B. R. Scheiber, University of Groningen, the Netherlands and Konrad Lorenz Research Station, Grünau, Austria, Brigitte M. Weiß, University of Tübingen, Germany and Konrad Lorenz Research Station, Grünau, Austria, Josef Hemetsberger, University of Vienna and Konrad Lorenz Research Station, Grünau, Austria, Kurt Kotrschal, University of Vienna and Konrad Lorenz Research Station, Grünau, Austria.
    pages   cm
Includes bibliographical references and index.
ISBN 978-0-521-82270-1 (hardback)
1. Greylag goose–Behavior.   2. Social behavior in animals.   I. Scheiber, Isabella B. R.
QL696.A52S652 2013
598.4´173–dc23
2012051610

ISBN 978-0-521-82270-1 Hardback
ISBN 978-1-108-81055-5 Paperback

Contents

List of contributors   *page* ix
Preface   xi
KURT KOTRSCHAL
Acknowledgements   xvi

Part I
Research background   1

1   Greylag geese: from general principles to the
Konrad Lorenz flock   3
JOSEF HEMETSBERGER, BRIGITTE M. WEIß AND
ISABELLA B. R. SCHEIBER
    1.1   Taxonomy   3
    1.2   Human–goose relationships   4
    1.3   Distribution   6
    1.4   Geese: general biology overview   6
    1.5   Greylag geese   8
    1.6   The greylag goose flock at the Konrad
       Lorenz Research Station (KLF) in Grünau,
       Upper Austria   10
    Summary   24
2   Goose research then and now   26
KATHARINA HIRSCHENHAUSER, HEIDI BUHROW, HELGA
FISCHER-MAMBLONA AND KURT KOTRSCHAL
    2.1   Konrad Lorenz and his geese   28
    2.2   Measuring behaviour then and now   29

2.3   A historical study of pair bonding behaviour
      in greylag geese   31
2.4   Today's perspective   37
Summary   40

Part II
From individual to clan   43

3   Individuals matter: personality   45
SIMONA KRALJ-FIŠER, JONATHAN NIALL DAISLEY AND KURT
KOTRSCHAL
3.1   (In)consistencies in within- and between-individual
      behavioural variation   46
3.2   Personality as a suite of correlated behaviours 49
3.3   Physiological processes underlying variation in
      personality types   50
3.4   Personality types and reproductive success:
      preliminary data   52
3.5   Experimental manipulations: enhanced yolk
      testosterone and its effect on behaviour   53
Summary   63
4   Maintenance of the monogamous pair bond   65
IULIA T. NEDELCU AND KATHARINA HIRSCHENHAUSER
4.1   Behavioural synchrony between pair partners   67
4.2   Androgen co-variation between pair partners: an
      indicator of mutual socio-sexual attraction?   77
4.3   Social modulation of hormonal co-variation   84
Summary   87
5   Alternative social and reproductive strategies   88
BRIGITTE M. WEIß
5.1   Individual life histories   89
5.2   Male–male pairs   90
5.3   Polygamous pair bonds   93
5.4   Alternative reproductive strategies   99
Summary   103
6   Beyond the pair bond: extended family bonds and
     female-centred clan formation   105
ISABELLA B. R. SCHEIBER AND BRIGITTE M. WEIß
6.1   Lineal (parent–offspring) bonds in the
      KLF greylag geese   108

6.2   Collateral (sibling) bonds among adult kin   114
Summary   117

Part III
Costs and benefits of social life   119

7   Causes and consequences of aggressive behaviour and
     dominance rank   121
     BRIGITTE M. WEIß
     7.1   Causes of aggressive behaviour and dominance
            rank in geese   122
     7.2   Consequences of aggressive behaviour and
            dominance rank   136
     Summary   141
8   The costs of sociality measured through heart
     rate modulation   142
     CLAUDIA A. F. WASCHER AND KURT KOTRSCHAL
     8.1   Implantation and transmitter technology   143
     8.2   Data collection   144
     8.3   Heart rate in geese: the overall picture   147
     8.4   Heart rate in the context of sociality   150
     Summary   154
9   'Tend and befriend': the importance of social allies
     in coping with social stress   156
     ISABELLA B. R. SCHEIBER
     9.1   Measuring social support in the KLF geese   159
     9.2   Active social support in greylag geese   161
     9.3   Passive social support in greylag geese   164
     9.4   Social support, one of the key factors in
            efficient stress management   170
     Summary   171
10   How to tell friend from foe: cognition
     in a complex society   172
     BRIGITTE M. WEIß, CHRISTIAN SCHLOEGL AND ISABELLA
     B. R. SCHEIBER
     10.1   Kin recognition   173
     10.2   Individual recognition   176
     10.3   Tracking and inferring relationships   180
     10.4   Conspecifics as sources of information   185
     Summary   187

Part IV
Lessons for vertebrate social life   189

11   The greylag goose as a model for vertebrate
     social complexity   191
     ISABELLA B. R. SCHEIBER, KURT KOTRSCHAL AND
     BRIGITTE M. WEIß
          11.1 The evolution of sociality   191
          11.2 Do birds have the brains for social complexity?   193
          11.3 Socially complex geese?   195
          11.4 The greylag goose as a model system for social
                complexity in vertebrates   199
          Summary   200

     References   202
     Index   233

Colour plates appear between pages 110 and 111.

Contributors

Heidi Buhrow (deceased 19 September 2009)

Jonathan N. Daisley
Heritage Environmental Ltd., Auchterarder, UK

Helga Fischer-Mamblona
Germany

Josef Hemetsberger
Konrad Lorenz Forschungsstelle für Ethologie, Grünau, Austria,
and Department of Behavioural Biology, The University of
Vienna, Austria

Katharina Hirschenhauser
Department of Behavioural Biology, The University of Vienna,
Austria

Kurt Kotrschal
Konrad Lorenz Forschungsstelle für Ethologie, Grünau, Austria,
and Department of Behavioural Biology, The University of
Vienna, Austria

Simona Kralj-Fišer
Institute of Biology, Scientific Research Centre of the Slovenian
Academy of Sciences and Arts, Ljubljana, Slovenia

Iulia Nedelcu
Department of Behavioural Biology, The University of Vienna,
Austria

Isabella B. R. Scheiber
Centre for Ecological and Evolutionary Studies, Behavioural Ecology and Self-organization Group, University of Groningen, The Netherlands

Christian Schloegl
German Primate Centre, Cognitive Ethology Laboratory, Göttingen, Germany

Claudia A. F. Wascher
Konrad Lorenz Research Station, Grünau, Austria

Brigitte M. Weiß
Institute for Evolution and Ecology, University of Tübingen, Germany

Preface
The social life of greylag geese: patterns, mechanisims and evolutionary function in an avian model system

KURT KOTRSCHAL

A book on the social fabric of greylag goose communities would be worth writing and interesting to read simply by virtue of the historical role played by these geese in the development of behavioural biology. Geese were one of the favourite subjects of Konrad Lorenz, one of the major founders of ethology (Tinbergen 1963). But there is much more to geese than this historical aspect. Konrad Lorenz' appreciation of the social complexities (defined by long-term dyadic, valuable and mutual relationships) of greylags and other geese was not all based on systematic data collection (Lorenz 1988). He was a keen observer and had a remarkable mental ability to identify, analyse and compare patterns. Therefore, much of his evidence was truly 'anecdotal'. This approach has certainly changed over time, not least because the standards of data gathering in science have definitely become more rigorous. This will be illustrated in the second chapter of this book.

In our book, we summarise more than 20 years of observational and experimental work on the 'Lorenzian geese' at the Konrad Lorenz Forschungsstelle (Konrad Lorenz Research Station; KLF) in Grünau, Austria. From the time of Konrad Lorenz onwards, this field station has been located in a picturesque valley of the Northern Alps. Substantial progress has been made with the semi-tame, free-roaming flock of geese living there. Several dozen colleagues and students, as well as a considerable number of volunteers, have contributed to our results over the past two decades, and we are very grateful to them all.

Our recent findings indicate that Konrad Lorenz underestimated, rather than overestimated, the complexity of the social life of geese. This book is a synthesis of social organisation in geese, mainly based on our own results, but embedded in the causal and functional knowledge acquired by others and framed by contemporary biological theory. Although this is a monograph on goose sociality, it is also relevant as a data point for comparing the structure and functions of social organisation in birds and mammals.

Jane Goodall's (1986) ground-breaking description of the social complexity and skills of wild chimpanzees motivated many biologists to come out of the laboratory and into the field. Konrad Lorenz was deeply impressed by her observational approach, which was just as keen but more systematic than his own. When they met, they enthusiastically agreed on the immense value of long-term studies on the social behaviour of animals, whether geese or chimps. From a present-day perspective we could not agree more. It took a while, however, before substantial scientific results started to be produced.

To an even greater extent than Konrad Lorenz, Jane Goodall opened a new window in animal behaviour research, which permitted animals to be approached as individuals who can think and have feelings. After decades of an extremely Descartian, automaton-like view of animals, which also prevailed in ethology, it again became possible to investigate cognition and emotions in animals without the risk of one's work not being taken seriously (Panksepp 1998; Aureli & Schaffner 2002; Aureli & Whiten 2003). In some instances it even became acceptable for scientists to become involved socially with their experimental animals; for example, by personally hand-rearing greylag goslings in order to obtain trusting partners for experimental work (Hemetsberger *et al.* 2010). Women such as Jane Goodall, Irene Pepperberg and Sue Savage-Rumbaugh demonstrated that rigorous scientific methods and compassion and empathy for one's study subjects do not have to be mutually exclusive. Konrad Lorenz is still the prime male example of a researcher having empathy with 'his' animals to such an extent that he was always opposed to invasive work with geese, while remaining a keen observer of their behaviour.

Ever since the time of Darwin (1872), it has been increasingly appreciated that the principles of social complexity may not be exclusive to human societies, and since the publication of Wilson's textbook (1975a), the functional rules governing social systems have become apparent. Primarily, the 'social brain hypothesis' (Humphrey 1976; Byrne & Whiten 1988; Dunbar 1998), developed using primates, made the connection between social complexity and cognitive development. The belief in primate–mammalian cognitive supremacy was deeply rooted in the Darwinian continuum. Birds, in particular, were not seriously considered to be candidates for the study of intelligence and complex social systems. This was a consequence of the long-standing misconception that birds' forebrains consist mainly of striatal rather than pallial components (Edinger 1929), which doomed birds to be perceived as relatively instinctive and 'non-cognitive' creatures. This may even have been one of the reasons why birds became the prime research models in classical ethology, with its traditional focus on the instinctive, stimulus–response components of decision making, although this may have been different in disciplines with a different focus, such as comparative psychology or cognition research. Today we know that bird forebrains, although lacking layered and columnar structures, still feature a similar proportion of pallial components as mammals (Güntürkün 2005; Iwaniuk & Hurd 2005; Jarvis *et al.* 2005), and that corvids rival apes in relative brain size (Emery & Clayton 2004). Even the primatologists finally seem to accept that birds are capable of top-rate cognitive performance.

In fact, we can learn a great deal about the conditions and constraints for the development of complex social systems through comparison with phylogenetically distant taxa. Apes are in some respects strikingly similar to humans, simply because our ancestry diverged only about 6 million years ago. Hence, a substantial amount of common social dispositions within the apes may be due to evolutionary inertia. It would be surprising to find fundamental human traits that are not also basically represented in chimps. To explain the selection pressures acting towards the evolution of social complexity and intelligence, comparisons with birds may be even more relevant than those with chimpanzees, because the common ancestor of apes and birds was probably a generalised reptile living in the late Palaeozoic, some 280–300 million years ago (Jarvis *et al.* 2005). Therefore, if common patterns are found in geese and primates, these are very unlikely to be due to direct common descent ('homology'), especially as our common reptile ancestor was probably not a genius at social cognition. Hence, if such close similarities are found in primates and in birds, simple phylogenetic inertia is probably not the answer. Evolutionary constraints of a conservative central nervous system may have had a role in the shaping of parallel structures in birds and mammals, for example the similarities in the expression of individual behavioural phenotypes ('personality'; see Chapter 3). In general, however, parallels in psychological, behavioural and social structures must have formed due to similar selection pressures from ecological and/or social sources in an 'analogous' manner. Hence, a comparison between geese and primates may be revealing in pinpointing such common selection pressures.

The parallels between goose and mammalian social organisation are indeed striking. They include long-term dyadic and parent–offspring relationships, alliance formation, female bonding and emotional social support, among others. The main topics covered in this book could just as well appear as chapters in a book that discusses mammalian social systems.

There are three main sections to this book: on the biological and historical basics (Chapters 1 and 2); on social patterns (from individual to clan, Chapters 3–6); and on the causes and consequences of social organisation (Chapters 7–10). This means that we treat social patterns and social physiology as a single unit of conditions and constraints, which are congruent with evolutionary function. For example, the social patterns of the monogamous pair and the clan are the structural background for the expression of social support which, as a consequence, will affect individual social efficiency and, ultimately, reproductive success.

We cover goose individuality and personality (Chapter 3); stability and synchrony in the crucial long-term valuable partnership, the monogamous pair bond (Chapter 4); alternative reproductive strategies (Chapter 5); clan formation and extended family bonds (Chapter 6); causes and consequences of dominance and aggression (Chapter 7); the costs (Chapter 8) and benefits of social life achieved through a variety of patterns of social support (Chapter 9); and, finally, cognition in a complex social society (Chapter 10). The relevant biology of geese, as well as specific information about the greylag geese at the Konrad Lorenz Research Station are summarised first (Chapter 1), followed by a chapter on the historical role of geese in science (Chapter 2).

The parallels between the social organisation of birds and mammals, as described in this book, may still be surprising, as these two phyla have had independent evolutionary histories for approximately 300 million years (Benton & Donoghue 2007 and references therein). The motivation for summarising more than 20 years of goose research at Grünau came with the realisation that not only human and chimpanzee societies are similar, but that we can also find similarities in birds and very likely throughout all social vertebrates. Therefore, goose social patterns will be put into perspective with other bird and vertebrate social systems throughout the book and, in particular, in the final chapter (Chapter 11). A doctoral student once commented jokingly to Konrad Lorenz that: '*Gänse sind auch nur Menschen*' (Rost 2001: p. 192) or – loosely translated – '*Geese are only human too*'. Although today we would consider this too sweeping a statement, some similarities are indeed there, because sociality not only carries benefits but also entails costs, which can be quite high – and this is true for 'everybody', whether fish or human. Therefore this book should cater to anyone with an interest in social systems.

Acknowledgements

This book represents the concerted efforts of many people over more than two decades. The editors thank all colleagues, students, goose-raisers, civil servants, volunteers and staff of the Konrad Lorenz Research Station, who contributed through their work to the data presented in this book and to the maintenance and monitoring of the flock. Although nobody has been forgotten, there are too many to be named individually. We are deeply obliged to various people at the University of Vienna. These include, among many other colleagues from the Departments of Behavioural Biology and Cognitive Biology, **Thomas Bugnyar**, **John Dittami**, **Eva Millesi**, **Dagmar Rotter** and **Anna Schöbitz** for their regular scientific input, discussions, encouragement and 'social support'. We are indebted to **Erich Möstl**, **Rupert Palme** and **Sophie Rettenbacher-Riefler** (Institute of Biochemistry, Department of Natural Sciences, University of Veterinary Medicine Vienna, Austria). Their constant help in developing, improving and validating the technique of non-invasive steroid determination from droppings was invaluable. Similarly, we are grateful to everyone involved in the heart rate telemetry project: **Walter Arnold**, the Biotelemetry group and the Veterinary team, especially **Gerhard Fluch**, **Thomas Paumann**, **Franz Schober** and **Wolfgang Zenker** (Research Institute of Wildlife Ecology, University of Veterinary Medicine Vienna, Austria).

We also would like to express our most sincere gratitude to the external reviewers (listed in alphabetical order), who took the time to read through the drafts: **Ralph Bergmüller** (Department of Eco-Ethologie, University of Neuchâtel, Switzerland), **Peter Kappeler** (Behavioural Ecology and Sociobiology Unit, German Primate Centre and Department of Sociobiology/Anthropology, University of Göttingen, Germany), **Sonja Ludwig** (Game and Wildlife Conservation Trust,

Barnard Castle, UK), **Marta Manser** (Institute of Evolutionary Biology and Environmental Studies, University of Zurich, Switzerland), **Kees van Oers** (Netherlands Institute of Ecology, Wageningen, Netherlands), **Markus Öst** (ARONIA Coastal Zone Research Team Åbo Akademi University and Novia University of Applied Sciences, Ekenäs, Finland), **Maude Poisbleau** (Department of Biology, University of Antwerp, Belgium), **Jouke Prop** (Arctic Centre, University of Groningen, Netherlands), **Amanda Seed** (School of Psychology, University of St Andrews, UK), **Jennifer Elaine Smith** (Department of Ecology and Evolutionary Biology, University of California Los Angeles, USA) and **Vincent Viblanc** (Department of Ecology and Evolution, University of Lausanne and Department of Ecology, Physiology and Ethology, IPHC, CNRS-UdS, Strasbourg, France).

We are grateful to the **University of Vienna** for continuous funding and support as well as the permanent support of the **Verein der Förderer der Konrad Lorenz Forschungsstelle**, the **Herzog von Cumberland Stiftung** and the **Cumberland Game Park, Grünau**. Funding for the work presented in this book was further provided by the **Austrian Science Fund** (FWF Projects: P10483-BIO, P12472-BIO, P12914-BIO, P15766-B03, P18601-B17, P18744-B03, P20538-B17, P21489-B17 and R30-B03), the **Fürst Dietrichstein's sche Stiftung**, and the Swiss **Fondation Pierre Mercier pour la Science**.

We appreciate the help and patience of **Megan Waddington** (Assistant Editor) and **Martin Griffiths** (Commissioning Editor, Life Sciences) at Cambridge University Press.

Last, but not least, we would like to thank the **local authorities**, **community of Grünau**, and the former **Duke of Cumberland, Ernst August IV, Prinz von Hannover** and the present **Duke of Cumberland, Ernst August V, Prinz von Hannover** for hosting the KLF over the years. Our particular gratitude goes to **O. F. M. Hüthmayr, F. M. Lindner** and all current and former staff of the **Cumberland Game Park, Grünau** for their collaboration and support of the work conducted within the park and the frequent help also given outside – we could never have dealt with the masses of snow by ourselves!

Part I   Research background

JOSEF HEMETSBERGER, BRIGITTE M. WEIß AND
ISABELLA B. R. SCHEIBER

1

Greylag geese: from general principles to the Konrad Lorenz flock

Over many years, greylag geese (*Anser anser*) have inspired much long-term scientific research. Thanks to this continuing work, our knowledge of social organisation in birds has greatly improved. Before presenting the latest findings in greylag goose research over the remainder of this book, we therefore introduce the reader to the species – its taxonomic affiliation and geographical distribution – as well as providing information about goose biology in general, and greylag goose biology in particular. This book focuses on a remarkable greylag goose flock at the Konrad Lorenz Research Station (Konrad Lorenz Forschungsstelle, abbreviated as KLF) in Grünau (Upper Austria), where much of our understanding of greylag goose biology has come together over the last 40 years. The origins of this flock, at that location, date back to the time of the late Konrad Lorenz (1903–89). We describe the KLF and also summarise the kind of research questions that can be addressed using this semi-tame goose flock, pointing out its distinctive features relative to wild goose populations, which may be relevant in the interpretation and generality of our findings.

1.1 TAXONOMY

Together with the Galliformes, the order Anseriformes (waterfowl) belongs to one of the oldest lineages of modern (neognathous) birds.

The Social Life of Greylag Geese: Patterns, Mechanisms and Evolutionary Function in an Avian Model System, ed. I. B. R. Scheiber *et al.* Published by Cambridge University Press. © Cambridge University Press 2013.

Recent evidence suggests that they originated during the Cretaceous period (Clarke *et al.* 2005). Although waterfowl phylogeny is still partly unresolved, one extinct (Cnemiornithidae, New Zealand geese) and three extant families are recognised: the Anhimidae (screamers), Anseranatidae (with a single representative, the magpie goose, *Anseranas semipalmata*) and the Anatidae, which includes over 140 species of ducks, geese and swans.

Livezey (1996) divides the Anatidae into five subfamilies: the Dendrocygninae (whistling ducks and allies), Anserinae (true geese, swans and ducks), the monotypic Stictonettinae (with a single representative, the freckled duck, *Stictonetta naevosa*), Tadorninae (shelducks and allies) and Anatinae (surface-feeding ducks and allies).

The subfamily Anserinae is then subdivided into one extinct (Thambetochenini) and three extant tribes: the Cereopsini (with a single representative, the Cape Barren goose, *Cereopsis novaehollandiae*), the Anserini (true geese), and Cygnini (swans). Within the true geese, there are two extinct and 16 extant species in three genera: *Anser* (grey geese), *Chen* (white geese; nowadays usually considered as a subgenus of, and included in, *Anser*) and *Branta* (black geese).

The greylag goose (*Anser anser*), the largest and bulkiest of the grey geese (length 75–90 cm; mean mass: males 3.5 kg, females 3.0 kg; Beaman & Madge 1998), is the type species of the genus *Anser* and also the ancestor of the domestic goose (*A. a. domesticus*) in Europe and North America. It is probable that two isolated populations of greylag geese existed in the late Pleistocene: one in coastal south-western Europe and one in inland south-eastern Europe and Asia, evolving into the three extant subspecies: *A. a. sylvestris* (Iceland, Scotland, Norway), *A. a. rubrirostris* (south-eastern Europe and eastwards) and *A. a. anser* (between the other two subspecies), respectively. The nominate *A. a. anser* from central Europe may be considered as an intermediate between *sylvestris* and *rubrirostris* but, because of an overlap in bill length with *sylvestris*, north-western and central European populations are often treated as single subspecies, *anser*. There are several populations in Europe – some discrete, others overlapping – and most populations are at least partially migratory. The 'lag' part of the name 'greylag goose' is derived from its habit of being one of the last of the migratory geese to move south in the winter.

1.2 HUMAN–GOOSE RELATIONSHIPS

Geese and their eggs have been an important human food source for thousands of years, which may have been the reason why geese were among the first domesticated birds. They were fully domesticated approximately 3,000 years ago to provide meat, eggs and feathers (Todd 1996). Furthermore, it is known that geese were kept in ancient Egypt, approximately 4,500 years ago: in a burial site in Medum (Egypt), a fresco depicts greylag and red-breasted (*Branta ruficollis*) geese so perfectly that the birds must have been at least semi-tame (Burton & Risdon 1987; Todd 1996); the painting is included in *Meyers Blitz-Lexikon*, published in Leipzig (Meyers, 1932). Geese were also mentioned in two poems by Homer (see discussion in Pratt 1994). Apart from their value as food, humans have also used domesticated geese for other important tasks; because geese are naturally vigilant and domestic geese can be relatively aggressive, they have been used as 'watchdogs' as far back as 390 BC, when sacred Roman crested geese warned the Roman garrison of the attacking Gauls (Aicher 2001: p. 48). Geese also carried out this task at the Dumbarton grain whisky distillery from 1959 until the closing of the premises in 2002 (the 'Scotch Watch'; see Todd 1996). 'Weeder' geese have been used with great success in the USA since the 1950s to control and eradicate troublesome grasses and weeds in a variety of crops and plantings, such as strawberries and cotton plants.

As their traditional wintering grounds have been taken over by agriculture, a number of goose species have started to feed on farmland instead. This rich food supply seems to be a major reason for the increasing numbers of most Eurasian geese. They can cause serious damage to crops (Madsen *et al.* 1999; Gauthier *et al.* 2001; Bowler *et al.* 2005), making them unpopular with farmers. Recently, conflicts have also arisen between the increasing number of summer-staging geese and farmers, and damage compensation claims have risen not only for the wintering period but also during the summer (Feige *et al.* 2008). Therefore, in many European countries, hunting is often allowed in order to protect pastures, at least during certain times of the year (Madsen *et al.* 1999). Greylag geese, in particular, have been – and still are – hunted extensively over most of their range (see Bowler *et al.* 2005). The drainage of wetlands, in addition to human persecution facilitated by the accessibility of nesting sites, especially during moulting when they are flightless, has led to extensive local declines in many populations. For example, in the Netherlands, the greylag goose was a common breeding bird until the beginning of the sixteenth century. From then on, its numbers started to decline continuously until it disappeared as a regular breeding bird in the first half of the twentieth century (Feige *et al.* 2008). After reintroduction programmes in the 1960s and 1970s, and the construction of large nature reserves, the number of breeding pairs started to rise enormously, from an estimated 150 pairs in the 1970s to 25,000 pairs in 2005 (Voslamber *et al.* 2007; Feige *et al.* 2008). Data from the Netherlands reflect the overall population increase: northern European numbers have been growing quite dramatically (Kampp & Preuss 2005; Austin *et al.* 2007; Voslamber *et al.* 2007; Farago 2010; Fox *et al.* 2010; Mitchell *et al.* 2010), whereas the eastern European populations are more fragmented and relatively small, although they are also on the rise, at least in some countries. This is not true for *A. a. rubrirostris*, which is thought to have declined quite substantially in numbers because of intense habitat fragmentation and its susceptibility to hunting pressure (Madge & Burn 1988).

1.3 DISTRIBUTION

Geese are cosmopolitan, but are absent from continental Antarctica and some islands, with the highest concentration occurring in the Northern Hemisphere. Most species in Europe, Asia and North America are facultative or obligatory migratory, and some species may breed as far north as the Arctic Circle (up to 66° 33′ 44″ N). The brant goose (*B. bernicla*), for example, breeds farther north than any other goose species, with its breeding habitats being located in Siberia and along the northern coast of Alaska and western Canada. It winters in coastal areas of Europe, North America and Japan.

Both the breeding and the wintering ranges of the greylag goose tend to be more southerly than those of the other *Anser* species. It breeds throughout northern Eurasia: from Iceland, Great Britain, Scandinavia and the Netherlands, east across Russia, and south into Mongolia and northern China (Sibley & Monroe 1990). Birds from Iceland overwinter in Britain, where they join with the resident population. Scandinavian birds migrate south-west over Europe to winter mainly in France and the Iberian Peninsula, while central and southern European populations overwinter around the Mediterranean basin, including some lakes in northern Africa. Birds that breed in eastern Europe migrate to the Black and Caspian seas and into northern Iran and Iraq. Even farther east, there are wintering grounds in northern India into Myanmar as well as the lowlands of southern China and central Asia.

1.4 GEESE: GENERAL BIOLOGY OVERVIEW

As a group, geese are easily distinguishable from other waterfowl by their size, long necks, honking calls and sociable nature. Sexes of *Anser* and *Branta* are monomorphic, with males generally being slightly larger than females (Madge & Burn 1988). On the whole, they are medium-sized (e.g. lesser white-fronted goose, *A. erythropus*; Ross's goose, *A. rossii*; length: 53–66 cm) to large (e.g. Canada goose, *B. canadensis maxima*; length: 90–110 cm), aquatic and terrestrial herbivorous 'grazers' and strong swimmers (Beaman & Madge 1998). As geese assimilate only a small proportion of their plant food, they may spend up to 80% of the daylight hours feeding (Ogilvie & Pearson 1994) and might also feed at night (Ydenberg *et al.* 1984), thereby exposing themselves to predators quite extensively. Much of the remaining time is typically devoted to resting and preening, as well-ordered and oiled feathers are a prerequisite for keeping waterproof. Outside the breeding season, geese are highly gregarious and frequently feed and roost in large flocks, with some species congregating in groups of many tens of thousands during migration and on staging and winter grounds (Todd 1996). Geese migrate in their well-known chevron pattern ('V'-formation), often flying at high altitudes. They migrate southwards on traditional fly-ways in the autumn in large and dense flocks, and stop at well-established staging grounds such as the Neusiedlersee in eastern Austria. Here, one might find a total of 35,000 bean (*A. fabalis*), greater white-fronted (*A. albifrons flavirostris*) and greylag geese gathered together (relative to around 300 greylag breeding pairs; see Ogilvie & Pearson 1994; although the newest estimates are based on the assumption of more than 1,000 breeding pairs in 2012; M. Dworak, BirdLife Austria, personal communication). At their staging grounds they remain for several weeks in large numbers. In some species, the final destination for overwintering is often reached only after a cold spell has occurred (Beaman & Madge 1998). Arrival back at the nesting grounds of northern-breeding forms greatly depends upon weather conditions, so geese will wait out the cold weather at their stop-over sites until the breeding grounds are mostly snow-free (Madge & Burn 1988). Some species, such as the bar-headed goose (*A. indicus*), nest in dense colonies, where nests are as close as the pecking distances to neighbours will allow. In other species, such as the swan goose (*A. cygnoides*), several nests may be located in favoured areas; while other species, such as the bean goose, maintain nesting territories. Greylag geese, the most southerly breeding of all the 'grey' geese (Madge & Burn 1988), mainly nest in isolated pairs, but are known to sometimes nest in colonies with nests as close as 2 metres apart (Madge & Burn 1988; Todd 1996).

　　Geese are well known for their strong family ties; goose pairs are long-term monogamous and may sometimes remain bonded for life. They are mostly seasonal breeders with highly synchronised laying periods in arctic species. Nests are usually built near water on the ground in the open or in vegetation. Clutches usually have 4–7 eggs in *Anser* and *Branta*, but are smaller in high-latitude forms, where also no replacement clutches are produced. Eggs are laid at intervals of 1–2 days and are incubated solely by the female for 24–30 days. Upon hatching, the downy young are tended by both parents, with the exception of 'brooding' (i.e. keeping the goslings warm for the first few weeks after hatching, when they are still unable to maintain their body temperature, particularly during cold weather), which is almost exclusively done by the female. Newly hatched goslings respond to parents with 'vee-calls' and outstretched neck for greeting. The herbivorous, precocial young leave the nest and follow parents from the second day post-hatching. Fledging periods are relatively short in high-arctic breeders, which utilise the long arctic days and great abundance of food for fast growth. For example, in snow geese (*A. caerulescens*) the young are fledged in approximately 40 days, whereas temperate species take longer (e.g. around 70 days in Canada geese). Non-breeders form moulting flocks in the summer, when they become flightless, and are joined by breeders a few weeks later. Some species perform a moult migration, which is a move to safer waters for moulting (Madge & Burn 1988). Young stay with their parents after fledging, at least through their first autumn and winter, and in some species through the spring migration. Furthermore, they may reunite with their parents at the end of one or more subsequent unsuccessful breeding seasons. Geese mature sexually at 2–3 years of age but may form durable pair bonds well before sexual maturity. They may also maintain family bonds, particularly with female kin, for many years (Frigerio *et al.* 2001a; Waldeck *et al.* 2008; Anderholm *et al.* 2009a), as discussed later in the book (Chapter 6).

1.5 GREYLAG GEESE

As with the other 'grey' goose species, greylag geese are typically grazers in open countryside (Madge & Burn 1988). They consume a wide range of food (Voslamber *et al.* 2004) and seem to prefer to feed from croplands and grasslands rather than from natural food sources (van der Wal 1998). They do not favour any particular crop but are flexible in adjusting to yearly crop rotations, as they have a high degree of site fidelity (Rutschke 1997; Feige *et al.* 2008). Greylag geese are also quite flexible in choosing breeding sites, as long as they are close to easily accessible bodies of water: they nest in natural habitats such as reed-beds, marshy swamps and small islands, as well as along estuaries and lakes (Sibley & Monroe 1990), but also on meadows and pastures. They have adjusted their breeding behaviour to the cultivated landscape that exists today in many locations (Berndt & Busche 1991; Rutschke 1997; Kalchreuter 2000; Bauer *et al.* 2005; Feige *et al.* 2008).

Greylag geese seem a little less distrustful of humans than other goose species: reintroduced feral populations tolerate the close proximity of towns and villages (Madge & Burn 1988). They appear to withstand a certain level of human and other disturbances, such as grazing of cattle or mowing. Because of grazing by cattle, the plants tend to be much younger and more palatable, and are higher in protein, which seems to be beneficial to the geese (Owen 1990; Feige *et al.* 2008).

In greylag geese, lifelong monogamy is the rule, with males and females associating all year, even when sexually inactive (Lorenz 1991; Kotrschal *et al.* 2006). The reproductive output of waterfowl may vary considerably not only between, but also within, species (Johnson *et al.* 1992), and greylag geese are no exception. Depending on nest site and location, average hatching success, for instance, may vary from about 30% to 80% (Nilsson & Persson 1994; Kristiansen 1998). Determinants of breeding success include predation on eggs and goslings (Sargeant & Raveling 1992) by both mammalian and avian predators (e.g. red foxes, *Vulpes vulpes*: Kristiansen 1998; hooded crows, *Corvus corone cornix*: Kristiansen 1998; Zduniak 2006), and also weather conditions (Wright & Giles 1988) and other environmental factors. Outside the breeding season greylag geese, like most other geese, are highly gregarious, with strong pair and family bonds. Moreover, shortly after their young hatch, parents may sometimes join together, forming large groups that are able to defend all the offspring by mobbing and attacking predators (Todd 1996). The young of each year group fledge at approximately 60–70 days of age, at the same time as parents regain their ability to fly after a one-month moult of their wing and tail feathers. Fledged goslings remain with their parents until the next breeding season. Adult and first-winter survival rates range from 40% to 90%, depending on populations and their wintering grounds (Nilsson & Persson 1996; Frederiksen *et al.* 2004; Pistorius *et al.* 2007). Parents and their subadult offspring leave the wintering areas together in the spring, but the latter move elsewhere when their parents return to the breeding grounds.

When observing a greylag goose flock, one will find that the leading edge contains a higher proportion of young geese, as they move and feed faster than adults, pulling their parents with them (Ogilvie & Pearson 1994). On closer inspection, it is possible to identify families: the young with their respective male and female parents nearby. Within a greylag goose flock there is a stable – but not strictly linear – rank order, with individual ranks mainly being conditional upon bonding status (see Chapter 7): families dominate pairs, who – in turn – outrank singletons. This keeps the flock relatively peaceful – most of the time. One goose will move towards another with an outstretched neck and head held low. Should the encountered goose be subordinate, it will simply move out of the way. However, if the encountered goose is dominant, it may not give way and a fight could develop (see Chapter 7).

1.6 THE GREYLAG GOOSE FLOCK AT THE KONRAD LORENZ RESEARCH STATION (KLF) IN GRÜNAU, UPPER AUSTRIA

The greylag goose is the most-studied *Anser* species, with most of the studies being performed with semi-tame flocks or under feral conditions. As a scientific model species, greylags are well known as the birds that the late Konrad Lorenz, one of the founders of ethology, used for his major investigation of the behavioural phenomenon of imprinting (Lorenz 1935). In 1973 he was awarded the Nobel Price for Physiology and Medicine, together with Karl von Frisch and Nikolaas Tinbergen, 'for their discoveries concerning organisation and elicitation of individual and social behaviour patterns' (from a press release of the Karolinska Institute in 1973). Lorenz' contributions stem mainly from his findings on greylag goose behaviour. Among his many books, two are devoted to 'his' geese: *The Year of the Greylag Goose* (Lorenz *et al.* 1978; English translation: Lorenz *et al.* 1979), a popular science account; and *Here I am: Where are You?* (Lorenz 1988; English translation: Lorenz 1991), which includes the detailed greylag goose ethogram still in use today.

Konrad Lorenz and his co-workers carried out a 'longitudinal study' of the social developments in the long-lived greylag geese, which was initiated in 1950 in Buldern (North Rhine–Westphalia, Germany) and continued at the Max Planck Institute for Behavioral Physiology in Seewiesen (Bavaria, Germany) from 1956 onwards, until our flock was established at the KLF in Austria in June 1973. For this purpose, a total of 148 geese – including near-fledged, human-raised juvenile geese as well as pairs with offspring and moulting adults – were transferred from southern Germany to Austria. After 2 years, approximately 75 of those remained in Grünau instead of flying back to Germany or elsewhere (Hemetsberger 2002). These formed the ground stock for the present KLF flock, which has ranged in size from 110 to 180 individuals ever since, including about one-third of individuals raised by human foster parents. From 1973 to 2011 we have collected standardised life history data for a total of 982 geese. These comprise the adult individuals transferred from Seewiesen to Grünau in 1973, 12 immigrants to the flock, the hand-raised geese, and a total of 562 fledged goslings emerging from the KLF flock during this time (mean ± standard deviation (SD) 15 ± 10.7 per year, range 0–44). Of the goose-raised fledglings, 479 hatched in the valley around the research station, while 83 goslings hatched elsewhere, for instance the Traunsee, approximately 20 km to the west, or the Chiemsee in Bavaria, Germany, some 200 km to the west, but came to the flock in the autumn with their parents.

The flock is completely unrestrained throughout the year, but is supplemented with pellets and grain twice daily on the meadows around the research station, with low quantities from spring to autumn, and with sustaining amounts during the winter. These regular feedings and the fact that several water bodies do not freeze up in winter, and hence provide night roosts safe from terrestrial predators, keep the flock in the valley and accessible for research throughout the year. Accordingly, this flock has never adopted a migratory tradition, although once in a while it happens that geese emigrate/disperse, or 'strangers' immigrate into the valley. As in other populations, natural predation (mainly by red foxes) is common and may account for the loss of up to 10% of the adult flock per year (Hemetsberger 2001). Most predatory events occur on the nests during laying and breeding in March and April, which is the most important cause of the often male-biased sex ratio in the flock (50–65% of males over the years). Predation and the often harsh weather conditions in spring also cause high rates of pre-fledging mortality; while up to 100 goslings may hatch each year, gosling survival to fledging rarely exceeds 25%.

All geese are marked individually with a unique aluminium ring from the Vogelwarte Radolfzell (Institute for Ornithology, Germany) as well as with coloured leg bands, which are also coded for year of hatching, family affiliation, as well as whether the goose is hand-raised or goose-raised. This allows detailed information on individual life histories to be recorded, including survival and reproductive success. The mean age of the geese in the KLF flock is 7.5 years, but some individuals live well beyond 20 years of age. The greatest recorded age was attained by a male called *Herr Viel*, who was almost 27 years old when he died in 1993. Like *Herr Viel*, all individuals have a unique name, which is a much easier way of referring to and keeping track of an individual than its colour combination. These names are written in italics throughout the book. As a note of caution, however, an individual's sex should not be deduced from its name, as sexes were not always apparent (from morphology, behaviour and/or genetic sexing) when the birds were marked and named.

1.6.1 The Valley of the Geese

The home base of the flock is the Konrad Lorenz Research Station (47° 48' 50" N, 13° 56' 51" E), which is located in an Alpine valley in central Austria, some 5 km south of the village of Grünau im Almtal (Fig. 1.1). From there the valley extends 13 km towards its dead end in the south, closed off by an Alpine range rising to more than 2,000 metres above sea level. A meandering river, the Alm, finds its origin in a picturesque lake, the Almsee, which is located at the southern end of the valley, approximately 8 km from the KLF. The research station itself is housed in a building dating from 1779, which was originally a mill (Fig. 1.2), and is surrounded by meadows, three ponds, and passed by the river Alm. The geese roam freely in the valley between the KLF and the lake, which they began to choose as their night roost soon after being transferred from Seewiesen, and have continued to do so until the present. Initially, geese also used to breed at the lake, either on a small island or using the marshes and reeds on the lake shore. Over time, most of the breeding has relocated to the relatively safe breeding huts in Oberganslbach, about 7 km north of the lake, or the local game park. The lake, however, remained popular among our greylags for another purpose: approximately half of the flock, particularly the non-breeders of the year, moult in the relative safety of these marshes, whereas families usually remain in Oberganslbach. Between the research station and Oberganslbach, the Cumberland Game Park, with its streams and ponds, opens its gates to visitors all year round and also hosts many of the KLF geese throughout the day.

Oberganslbach (OGB; literally translated as 'upper goose creek', a name coined by Lorenz himself) was artificially created as a breeding area for the geese in 1973. Part of the river was rerouted through a deforested area of approximately 190 × 90 m, which now consists of several creeks and a large pond that hosts three small islands (Fig. 1.3). Geese breed in open nests on these islands, in natural nests along the streams, or in one of the 21 relatively predator-safe nest boxes

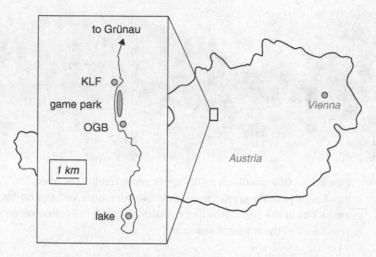

Figure 1.1   The valley of the river Alm in central Austria and the locations that are regularly used by the KLF greylag goose flock (KLF – Konrad Lorenz Research Station; game park – Cumberland Game Park; OGB – Oberganslbach).

Figure 1.2   The Konrad Lorenz Research Station, an old mill house, in Grünau, Upper Austria. © B. M. Weiß. A full-colour version is included in the colour plate section.

Figure 1.3   Oberganslbach (OGB; 'upper goose creek'), the main breeding area of the greylag geese. Small wooden huts for hand-raising are located at the edge of the forest. © B. M. Weiß. A full-colour version is included in the colour plate section.

provided by the KLF (Fig. 1.4), which are located at the research station and in the Cumberland Game Park as well as in OGB. The majority of families raise their young in OGB, and it is also the location where goslings are hand-raised. The 'luxury woodland accommodation' established for this purpose includes three wooden huts (approximately 2 × 3 m each), which in recent years have been equipped with solar panels for electricity, individual privies in the forest, and two wells for drinking water. Socially involved hand-raising follows the Lorenzian tradition even today (Hemetsberger *et al.* 2010).

1.6.2 The hand-raising tradition

Eggs for hand-raising are usually collected from abandoned nests in the KLF flock and from different goose populations throughout Europe. The latter ensures that genetic diversity is maintained in this rather isolated population and reduces the need to interfere in the breeding activity of the local flock. Eggs are incubated and hatched in a commercial incubator at the KLF. To mimic natural family size and ensure proper socialisation, we aim to raise sibling groups of 3–7 goslings. If necessary, we assemble a sibling group from different clutches, as goslings raised together seem to perceive each other as family (Kalas 1977). Accordingly, it is possible that not all individuals of one hand-raised group are genetically closely related or are exactly the same age (age difference at most 7 days). Goslings are marked individually immediately after hatching and are given to their human foster parent a few hours later, when the goslings have recovered

Figure 1.4   Inside (*upper*) and outside (*lower*) view of the predator-safe
nest boxes located in the ponds throughout the valley. © B. M. Weiß.

from the hatching process and have shed the thin horny sheaths that
covered the down feathers while in the egg. The young 'family' stays
close to the research station for the first 2–3 days and then moves to
OGB. Here the goose-raisers spend approximately 3 months with their
goslings, day and night without leaving them, until they fledge. The
human–goose families undertake daily excursions, as goose families
do, following the same spatio-temporal patterns. The raisers usually
stay close to goose families as well as to the rest of the flock to acquaint
the hand-raised goslings with other geese and to ensure 'proper' goose
behaviour throughout their lives. This is also a major reason why

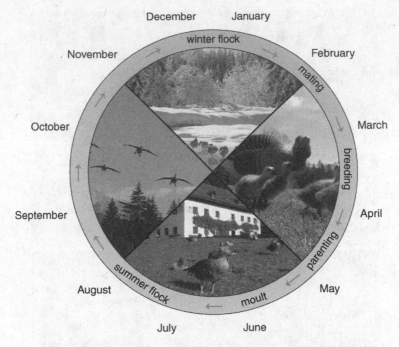

Figure 1.5   The annual cycle of the Grünau goose flock. Modified from Kotrschal (1995); photos © B. M. Weiß. A full-colour version is included in the colour plate section.

goslings at the KLF are always hand-raised in sibling groups, as goslings raised singly by human foster parents might also become sexually mis-printed. This 'socially involved hand-raising' keeps the flock relatively tame and thereby accessible to humans. Resulting from the careful hand-raising tradition, hand-raised geese show life histories similar to the goose-raised geese, but have reduced stress responses when inter-acting with humans in an experimental setting (Hemetsberger *et al.* 2010). Hand-raised goslings are amenable to behavioural experiments which would hardly be possible with goose-raised individuals.

1.6.3 A typical goose day and year

The importance and use of different locations in the valley changes throughout the day and over the year. In summer, after the flock has re-established itself around the research station following moulting (Fig. 1.5), a typical 'goose day' in Grünau would be as follows: the geese spend the night at the Almsee, flying to the research station shortly after dawn. They gather in front of the house and wait for

Figure 1.6 Great expectations: the flock assembles in front of the research station in anticipation of the afternoon feed. © B. M. Weiß. A full-colour version is included in the colour plate section.

their morning feed. Thereafter, they spend the morning around the research station bathing and resting, interrupted by a grazing bout approximately every 2 hours. Around noon they move down to the river or hop over into the game park, where they may snatch a piece of bread or a peanut from the park visitors. Late in the afternoon they return for their afternoon feed (Fig. 1.6), spend the early evening around the station, and at dusk flock together to fly back to the lake (Fig. 1.7a) after an elaborate pre-flight ceremony (Lorenz 1991). This spatio-temporal pattern remains similar throughout autumn, but the time spent in the game park gradually decreases until winter, when the geese come from the Almsee in the morning, and then spend the entire day around the research station or nearby at the river. They no longer visit the game park, as visitors are scarce and there is no chance of some extra goodies. Also in winter they return to the lake at dusk (Fig. 1.7b). On very cold days, they spend their day in or along the river in front of the research station, and they often have to be 'invited' (i.e. called up from the river) for feeding and then immediately return to the water after being fed, because the running water of the river is still much warmer than the frozen ground. In late winter/early spring,

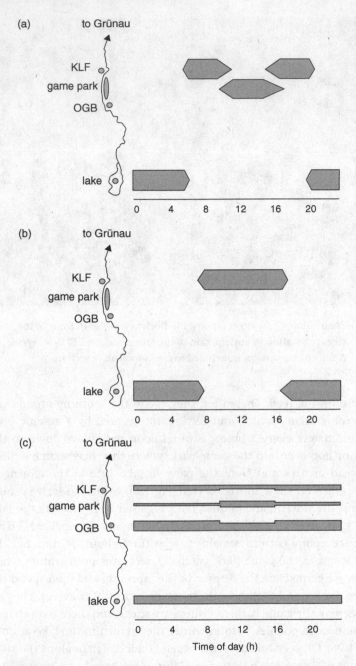

Figure 1.7 Daily use of the valley by the KLF goose flock in (*a*) late summer and autumn, (*b*) winter and (*c*) spring and early summer. *Grey bars* indicate the location of the flock at a given time, *bar thickness* indicates the proportion of the flock in a given location (in Fig.1.7c only).

agonistic interactions in the flock increase and the flock disintegrates and spreads out more over the valley, as pairs start visiting the game park and OGB in search of nest sites. Eventually, most activities shift from the KLF to OGB, which is now also used as a night roost by most breeders and some of the non-breeders (Fig. 1.7c). Only a few families prefer to spend the days and nights at the ponds of the KLF or the game park. Accordingly, feeding of the flock is shifted mainly from the KLF to OGB in spring and early summer. The remaining non-breeders move up to the lake, where they spend the days and nights, often well hidden in the bogs and reeds of the shore, until moult of the wing feathers is completed in early summer. From July onwards, juveniles and adults have (re-)gained their ability to fly and the activities of the flock shift back to the KLF once again.

1.6.4 Research at the KLF: routines, assets and constraints

Since the establishment of the flock in 1973, the life histories of all geese have been monitored closely, including records of each individual's ancestry, its survival, pair bonds and reproductive success. Since 1990, monitoring has further intensified and also includes records of the precise whereabouts of all individuals three times per week, and in spring a rigorous monitoring of nest sites every second day. New eggs are numbered consecutively, weighed, and length and width are measured (Hemetsberger 2001, 2002). After hatching, families are monitored daily to determine the number of young and to check on their proper development. Close to fledging at approximately 8 weeks of age, goslings are caught by hand or in a stationary aviary trap for ringing and taking body measurements such as mass and tarsus length. On this occasion a blood sample is also taken, mainly for genetic determination of parentage. All of these data are pooled in a 'goose data library' that contains systematic records about each individual's hatching site and date, its parents, whether it was goose-raised or hand-raised, all pair bonds, survival, and reproductive data such as nest site location, number of eggs, hatching date and hatched and fledged goslings per year. These records build an extensive set of information available to all goose researchers at the KLF. They allow us to monitor the development of the flock, such as changes in the age structure (Fig. 1.8) or gradual shifts in the onset of the breeding season (Fig. 1.9), and form the foundation upon which much of the research at the KLF is built.

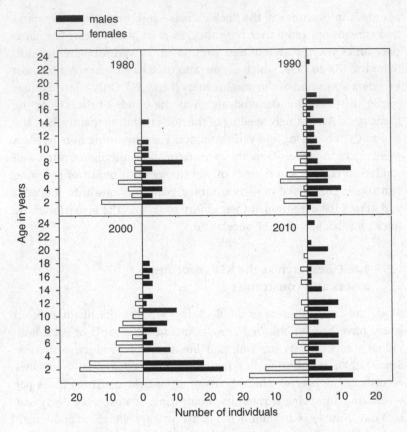

Figure 1.8   Age structure (*black bars*: number of males; *white bars*: number of females) of the KLF goose flock in the years 1980, 1990, 2000 and 2010.

Since the death of Lorenz in 1989, the main focus of investigations involving the geese was to understand the patterns of goose sociality and the mechanisms underlying it; specifically, how individuals in the flock manage to survive and be successful in a challenging social environment, as social stressors are among the most potent (DeVries *et al.* 2003). Data collection for these studies includes 'simple' behavioural observations, but also experimental manipulations and some measurements of physiological parameters. Due to the flock's high tolerance to human presence, behavioural data can be collected from a close distance without disturbing the geese. This was shown by heart rate recordings while approaching resting geese (Wascher *et al.* 2011).

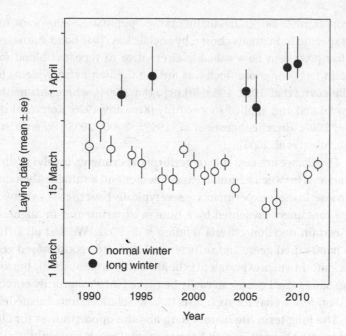

Figure 1.9   Onset of the breeding season as measured by the mean laying date from 1990 to 2012. Years following a typical winter are depicted as *white circles*; years following an exceptionally long winter as *black circles*.

In addition to observations under 'natural' conditions, experimental manipulations of various kinds are performed quite regularly. Behavioural experiments include the manipulation of food distribution or exposure to stressors, such as a predator model, usually a leashed dog, or the brief catching and immobilisation of a goose by holding it. Some experiments were conducted to manipulate or measure specific aspects of physiology, for example injecting eggs with testosterone for manipulation of the personality axis (Chapter 3) or implanting testosterone-filled tubes to study testosterone-sensitive behaviours of male geese (Frigerio *et al.* 2004a), to the point of inserting abdominal transmitters for measuring heart rate and body temperature in different social situations (Wascher *et al.* 2008a; see also Chapters 3 and 8).

However, the repeated catching or confinement of flock members is difficult and usually feasible only with very tame individuals (e.g. Kralj-Fišer *et al.* 2007, 2010a; see Chapter 3). Fortunately, much can be learned about goose physiology non-invasively by measuring steroid hormones such as testosterone, oestrogen and corticosterone in droppings. As our individually marked population allows for sampling specific individuals chosen by social class, pair bond status, etc., this has proved to be a valuable alternative to repeated blood sampling. In fact, our goose flock was among the first avian systems (see also Bercovitz *et al.* 1982, 1988; Kikuchi *et al.* 1994), where this method was validated and applied successfully (Krawany 1996; Kotrschal *et al.* 1998a, 2000; Hirschenhauser *et al.* 1999a, 2000, 2005; Frigerio *et al.* 2004b; Möstl *et al.* 2005).

Finally, we are also able to perform experiments involving direct interactions between a human experimenter and a sufficiently habituated goose. In these experiments, geese typically have to choose between baited containers presented by a human experimenter in alignment with certain decision criteria (Pfeffer *et al.* 2002; Weiß *et al.* 2010a). Both hand-raised geese and sufficiently habituated goose-raised geese participate in such experiments (B. M. Weiß, unpublished), but only young hand-raised geese can easily be coaxed into voluntarily entering closed-off test arenas for such tests (I. B. R. Scheiber, unpublished).

The long-term life history data and the opportunities for close behavioural observations and experimental work certainly provide extraordinary opportunities to study social complexity, physiology and cognition in a social bird. However, there are unresolved questions, particularly pertaining to greylag goose ecology, that we can address only to a limited extent and/or that may not be fully transferable to other populations of greylag geese. Like many other flocks of geese, our flock is fully free-ranging throughout the year, and is therefore exposed to abiotic factors such as temperature and precipitation as well as certain biotic factors, such as predation. Nonetheless, the KLF manages the flock in certain respects. First and foremost, our flock is supplied with food on a daily basis and, although there is competition around the feeders, all are provided with food of equal quality. Hence, food availability may not be considered a limiting factor, and the widely varying reproductive success in this flock (Chapter 5) may not be influenced primarily by individual differences in foraging efficiency. However, relationships between feeding and social hierarchy do exist even in this food-provisioned flock (Kotrschal & Hemetsberger 1995; Chapter 9), and in truly wild flocks it is well documented that dominance status influences the amount and quality of food an individual can obtain (Black *et al.* 1992, 2007; Kotrschal & Hemetsberger 1995; Stahl *et al.* 2001). For instance, high-ranking individuals can be found at the leading edge of the flock, where food is very probably of better quality, whereas pairs and singletons are more likely to be found in the centre of the flock or at other edges (Black *et al.* 1992; see also Box 9.2 in Black *et al.* 2007: p. 133).

In addition to feeding, we sometimes intervene in a way that affects fitness factors. In greylag geese, female mortality during incubation and losses due to egg predation can be substantial (Witkowski 1983; Osiejuk & Kuczynski 2007). At the KLF, geese are not only offered the preferred predator-safe nest boxes but also, if we find the nests in time, we remove natural nests from locations where predation on the breeding female is highly likely to occur. Therefore, the hatching success of our geese may be higher than in many natural greylag populations (Newton & Kerbes 1974; Kristiansen 1998). This higher hatching success, however, is counterbalanced by elevated constraints on pre-fledging survival posed by life in an Alpine valley, particularly in years with high red fox densities and/or particularly wet and cold weather conditions.

Furthermore, geese are known to lay eggs in the nests of others (Weigmann & Lamprecht 1991), and our geese are no exception (Chapter 5). Some clutches may contain more than ten eggs from various females and are likely to be abandoned before the onset of or early during incubation. We therefore remove eggs in nests that contain 9–10 eggs or more. The removed eggs are either discarded or hatched in an incubator for hand-raising, but are never foisted on other females. In addition, if an insufficient number of eggs from outside the population are available for hand-raising, we remove some eggs, or even complete clutches, from our own breeding females according to certain criteria. All of these clutch manipulations are carefully recorded and can thus be taken into consideration, but will necessarily reduce the amount of data available for estimating parameters related to reproductive success. In general, our flock is an excellent research model for behaviour patterns and mechanisms and for the behaviour–physiology interface. However, data on reproductive parameters to be used in ultimate (functional)-level questions will need to be carefully reviewed and selected before use.

Finally, mortality in geese has been suggested to be highest during migratory flights, particularly for subadults in their first year or during bad weather conditions (barnacle geese, *B. leucopsis*: Owen & Black 1989; white-fronted geese: Fox *et al.* 2003; but see Gauthier *et al.* 2001 for greater snow geese, *A. c. atlantica*). Also, as already discussed above, hunting has been shown to have a large effect on mortality in geese (Gauthier *et al.* 2001; Bowler *et al.* 2005). Again, these two factors are irrelevant for the KLF flock. Although our geese are officially allowed to be hunted under Austrian law, the hunters in the valley spare them without exception. This, of course, does not extend to geese that leave the valley and, indeed, occasional dispersers may get shot in other parts of Austria or Europe. Consequently, the first-year survival of individuals in the flock may be higher than in fully natural goose populations, although it should be noted that other greylag goose populations, particularly in urban areas, are also non-migratory and not subject to hunting.

At times, the unique situation of the KLF geese has been criticised for being 'artificial' or 'unnatural', and the conditions under which this study population live are indeed special. But any free-living population of geese, or animals in general, faces certain particular conditions that differ from the conditions encountered by conspecifics in other populations. For most biological questions, variation caused by such special conditions will not render the results obtained in one population inapplicable to others when being considered in the interpretation of results. In fact, adaptations of populations to the local conditions are an interesting study topic in their own right.

SUMMARY

This is an introduction to geese in general, i.e. their classification, general biology, and relationship with humans. In addition, we explain greylag goose biology to the reader, and provide an insight into the Lorenzian goose flock at the Konrad Lorenz Research Station in Grünau, Upper Austria. We describe a typical goose day and year, as experienced here, as well as the research possibilities; describe the often-cited hand-raising tradition, which dates back to Lorenz; and then go into the benefits and constraints one has to accept when working with this flock. The KLF flock of greylag geese is indeed a special study population. The flock is well suited for behavioural studies from a close distance, where one's attention is focused on the individual rather than the greylag goose as a species. Habituation to humans allows for experimental manipulation to a certain extent, particularly in answering questions pertaining to the physiological mechanisms underlying sociality. To a more limited extent, the flock is also suitable for addressing ecological and evolutionary questions. The knowledge gained from studies performed with this flock over the last two decades has substantially contributed to our understanding of social complexity in the animal world. The major findings on the social life of geese obtained with this flock in the two decades since Lorenz died are presented and synthesised in this book.

KATHARINA HIRSCHENHAUSER, HEIDI BUHROW†, HELGA
FISCHER-MAMBLONA AND KURT KOTRSCHAL

2
_____

Goose research then and now

Geese are a good source of food and produce excellent down, but geese and science? Everyone has heard of a 'silly goose', but geese, and greylag geese in particular, played a central role when Konrad Lorenz and others developed the principles of ethology in the early 1930s (Lorenz 1935). In fact, birds have always been central to ethology. Most of its early founders such as Oskar Heinroth, Robert Hinde, Klaus Immelmann, Gustav Kramer, Daniel Lehrman, Konrad Lorenz, William Thorpe, Niko Tinbergen, Charles Otis Whitman, William Thorpe and many others were excellent biologists with a special aptitude for theory development. However, at heart, they were all enthusiastic and devoted 'birders'. To this day, the attendees at major ornithological and ethological meetings overlap to a great extent.

In contrast to many nocturnal and often olfactorily orientated mammals, most birds are active during the day and they are mainly visually and acoustically orientated. Their diversity of lifestyles, and even their social organisation, is remarkable. Bird biology, in particular their visual and acoustic signals, is highly accessible to the human observer. Birds are species-rich, and therefore comprise an excellent group for comparative work (Lorenz 1939a). It is hardly surprising that many basic principles of ethology were developed in birds. This raises the question of how much the development of ethological theory was biased by the use of avian model species. The heated debates of the early 1960s between the mainly European behavioural biologists and

The Social Life of Greylag Geese: Patterns, Mechanisms and Evolutionary Function in an Avian Model System, ed. I. B. R. Scheiber *et al.* Published by Cambridge University Press. © Cambridge University Press 2013.
† Deceased 19 September 2009. This chapter is dedicated to her memory.

American experimental psychologists may have also been due to their different model organisms, songbirds in the case of the ethologists and mostly rats – but also pigeons – in the case of the psychologists (Kotrschal *et al.* 2001).

Although other groups of vertebrates show many of the features found in birds, it seems that the latter are simply one of the easiest groups to work with. Teleost fish, for example, are a highly diverse group that equal birds in many ways, i.e. in the sensory (Kotrschal & Palzenberger 1992; Kotrschal 1998; Dobberfuhl *et al.* 2005; Sovrano *et al.* 2007), ecological (Fricke 1975a; Kotrschal *et al.* 1998b; Ito *et al.* 2007), cognitive (Fricke 1975b; Höjesjö *et al.* 1998; Oliveira *et al.* 1998; Bshary *et al.* 2002; Grosenick *et al.* 2007) and social domains (Taborsky 1984; Milinski 1987; Höjesjö *et al.* 1998; Ward & Hart 2003; Hirschenhauser *et al.* 2004; Filby *et al.* 2010; see also review in Keenleyside 1991). Similar to birds (see also Chapter 11), at least one group of teleost fish, the ray-finned fishes (Actinopterygii) have evolved a kind of pallium, i.e. telencephalic nuclear masses with similar functions to the laminar cortex of mammals (Ito & Yamamoto 2009). Although scientists have recently started to appreciate the great potential of making their studies 'fishier', working with fish in the wild still needs much more effort and gear, as fish watchers have to immerse themselves in a medium that they were not designed for. Observing birds often requires little more than a pair of binoculars. Hence, there are many more bird watchers than fish watchers around.

One may also speculate whether it is easier to observe birds because their behaviour may seem more stereotyped (consisting of chains of clearly identifiable action patterns) than, for example, the behaviour of the primarily olfaction-orientated mammals. Less than 20 hours of observation with minimal guidance is usually sufficient for a student to produce a reasonably complete ethogram; i.e. a catalogue of greylag behaviour. The reason is that birds, as with their human observers, are visually and acoustically orientated, and their communication predominantly uses these two channels (but see Birkhead 2012 and references therein). Mammals, on the other hand, are more difficult, as many species use their olfactory sense for communication. While the ethogram of a laboratory mouse might look simple at first glance, mice communicate through smell and produce sounds in the hypersonic range, bypassing our eyes, ears and noses. Therefore, it is not surprising that many principles of ethology were first developed in birds.

Figure 2.1   Konrad Lorenz with a greylag goose family in
Oberganslbach, about 1975. © S. Kalas.

2.1 KONRAD LORENZ AND HIS GEESE

Geese were among the many species of waterfowl kept and observed
by young Lorenz in preparation for one of his pioneering comparative studies on the Anatidae (ducks, geese and swans; Lorenz 1941).
Here he demonstrated that the study of behaviours is as suited to
taxonomy and comparative work as are bodily structures (as already
shown by Whitman 1899 and Heinroth 1911) and therefore must be
species-specific, 'innate' and amenable to selection. The motor coordination required for egg retrieval in greylag geese (Lorenz & Tinbergen
1938) served as the key behaviour for one of the central elements
of this theory, '*Erbkoordination*' ('fixed action pattern'). Greylag geese
also played a pivotal role in early studies of imprinting (Schutz 1965;
Hess 1973), and the impressive man with the grey beard followed by
goslings is still one of the most popular icons of ethology worldwide
(Fig. 2.1).

   Lorenz himself was not entirely happy with this work, as the
geese that he and his friend Niko Tinbergen observed during the
beautiful summer of 1934 at Lorenz' home in Altenberg were greylag
goose × domestic goose hybrids. Lorenz always proposed that domestication and hybridisation are major causes of the 'degeneration' of
the genetics behind the expression of 'social instincts' in animals,

including humans (Lorenz 1940, 1973; Kotrschal *et al.* 2001). The political implications of these ideas during the turbulent times of Nazi rule and the significance of Konrad Lorenz' work before and during World War II have been the subject of much debate (Föger & Taschwer 2001). Initially, he remained unconvinced that their results would be representative of the 'noble savage', i.e. truly wild greylag geese. Later replication of these observations with greylag geese at Seewiesen showed that the behavioural development of wild geese was actually similar to that of domestic geese. Greylag geese remained one of Lorenz' animal companion species and research models for life (see Lorenz & Tinbergen 1938; Lorenz 1991).

At the Konrad Lorenz Research Station in Grünau, Austria, a few elderly male greylag geese that had met the late Konrad Lorenz were still around until recently; the last individuals from that era vanished in 2011. The 'Lorenzian' flock after Lorenz, more than ever before, supports behavioural research at the interface between mechanisms and functions, and this book summarises this research in the various chapters. Geese have been valuable research models elsewhere as well (Lamprecht 1986a, 1986b; Cooke *et al.* 1995; Black 1996; Loonen 1997; Black *et al.* 2007).

2.2 MEASURING BEHAVIOUR THEN AND NOW

Classical ethology was based on observations and was purely descriptive. This has changed dramatically between then and now. Modern ethology requires clear and testable hypotheses and predictions. We need to test them using quantitative data and often use sophisticated multivariate statistics to analyse the observed patterns. Just as an anatomist uses millimetres to compare the length of a bone from a set of individuals, the ethologist needs measurable units for comparing the expression of behaviour among a group of individuals. Nowadays we also have methods to measure behaviour with a minimum of disturbance caused by the presence of an observer, either by using remote video-recordings and photo-traps or fully automated data loggers. Advanced radio-tracking, geographical satellite navigation technology and an interdisciplinary combination of methods provide tools to collect information on an animal's movement patterns in the wild – even without seeing the animal itself (Wikelski *et al.* 2007; Hebblewhite & Haydon 2010). For example, the development of non-invasive DNA-based methods for the genotyping and sexing of individuals as well as the measurement of steroid hormone derivates from hair, feathers or excreta has opened up new opportunities for remotely tracking wild animals (Creel et al. 2003; Bortolotti et al. 2008; Segelbacher & Höglund 2009; Jacob et al. 2010). The near future will bring more exciting data on animal movements using a new generation of technological support (Sakamoto et al. 2009; Nathan et al. 2012).

With regard to the geese at the KLF, we now have a great opportunity for experimental work in the field with this free-roaming flock. One might call it a 'semi-natural setting'; while the opportunities for experimental work within the birds' social environment, as well as detailed studies of individual life histories (where climate and predation pressure remain unpredictable factors) are indeed exceptional. For instance, subcutaneously implanted transponders first enabled automated data collection on individual competence in initiating the use of new food resources (Pfeffer et al. 2002). Studies on the social energetics of individual greylag geese based on advanced heart rate telemetry can be found in Chapter 8.

Despite all these technical advances, the basis of the quantitative study of behaviour is (still) the 'classical ethogram'. We can only start assessing a variable after having clearly defined what we are measuring. As Tinbergen (1963) said: 'We need to define a form before we can study its function, causation and purpose.' Lorenz and his colleagues made exemplary efforts in producing a detailed ethogram of the behavioural repertoire in greylag geese (summarised in Lorenz 1991). These, and descriptive reports on the social interactions between the individuals of the goose flock (even if anecdotal), are the reason why geese are now considered a model species for studying the basic principles of social organisation in vertebrates in general.

In 1950, Konrad Lorenz started to establish a flock of greylag geese in Buldern at the Max Planck Institute for Behavioral Physiology, which was transferred to a location at the Bavarian Eßsee (called 'Seewiesen') by Wolfgang Schleidt in 1958. The geese were hand-raised by humans and some wild-caught geese were integrated into the flock (see also Chapter 1). Research focus was then on long-term observations of the social structure of this free-living population of tame greylag geese to study the mechanisms and functions of partnerships and sociality. In the following sections, we report on one specific set of data collected by Heidi Buhrow and Helga Fischer-Mamblona at Seewiesen in the 1960s. From a present-day reviewer's perspective, there are some problematic aspects of these data, as well as in the way they have been analysed. However, it is remarkable how the questions are still topical today and that the conclusions were close to what they would have been using modern data analyses. In fact, many pieces of the puzzle on the complexities of social behaviour of greylag geese were already there, albeit mainly based on observations and experience rather than experimental work and statistics. Due to the lack of adequate methods, the linkage of behaviour to the underlying hormonal and neuronal mechanisms was still in its infancy.

The data set presented here has never been published previously and could only be integrated here with the active help of Heidi Buhrow and Helga Fischer. This chapter pays tribute to the painstaking early work on geese done by Konrad Lorenz and his collaborators in Seewiesen. A wealth of data from this time is still unpublished and will probably remain so. The study presented here is included as a historical example of studies conducted 50 years ago. A number of details were no longer available to us or could not be recalled in enough detail by Heidi Buhrow and Helga Fischer, and therefore this chapter ends with a list of comments that a contemporary 'reviewer' might provide if this report was submitted for publication in a current ethological journal.

2.3 A HISTORICAL STUDY OF PAIR BONDING
BEHAVIOUR IN GREYLAG GEESE

Detailed documentation of individual life histories from the Lorenzian greylag geese at Seewiesen go back to the year 1954. These notes focused primarily on the relationships between individuals, with emphasis on mate choice, breeding success and the duration of relationships. Among the first questions that were intended to be answered from this data set were:

1. What are the behavioural elements of pair formation?
2. What characterises a pair bond?
3. How long does a pair bond between two individuals last?

From the data it soon became apparent that the social structure of a goose flock was more complex than the researchers had initially assumed. Rather than simply comparing the behaviour of different pair partners with each other, the variability of relationships between the partners appeared to be crucial. For example, varying numbers of partners (e.g. pairs and temporary trio formations) or divorces resulted in variable pair bond durations. It became clear that pair behaviour is more than the sum of the contributions of two individual partners.

Rather, the interaction produces 'emergent properties', as Konrad Lorenz stated repeatedly (Lorenz 1992).

In 1969, the pair bond structure of the Seewiesen goose flock was analysed for the first time, based on 15 years of continuous life history records from 61 males and 71 females. In addition to unpaired individuals, pairs were included in the analysis if continuous records for at least 4 years after hatching were available. Relevant behaviours for indicating an existing relationship between two (or more) individuals were defined as occurring in three contexts:

- in the context of affiliation (i.e. approaching, greeting, triumph ceremony, see Chapter 11)
- in a reproductive context, including sexual and parental behaviours
- during conflict (i.e. support of family members during agonistic interactions and rank order).

Based on these criteria, relationships were classified as 'primary partnerships' and 'secondary partnerships' (liaisons). Primary partners were geese that were paired with each other for at least 1 year and directed pair bond-associated behaviours, such as triumph ceremonies (Fischer 1965) towards one another. Relationships between secondary liaison partners lasted less than 1 year and pair bond-associated behaviours were not fully expressed or were not exclusively directed towards each other. Relationships with secondary partners were further divided into short-term and long-term liaisons. Short-term liaisons did not last longer than 1 week and were characterised by minor rates of partner-directed behaviours. Long-term liaisons lasted in the range of several weeks to months (in extreme cases for years, even if that included intervals of interruption) and the ('secondary') partners in such long-term liaisons performed behaviours mutually and in all contexts. 'Trio' formations were defined as continuously directing triumph ceremonies to more than one partner and the absence of secondary liaisons. In contrast to trios, secondary long-term liaisons were being attended repeatedly during mating seasons but were not obvious throughout the year.

Following these definitions, the first quantitative study on greylag goose relationships was conducted, based on 228 such social units. The aim was to describe the structure, frequency of occurrence, and biological efficiency of primary pair bonds and secondary liaisons (which would be 'extrapair activities' in contemporary terms). Remarkably, already in those early days of sociobiology and eco-ethology, Konrad

Lorenz and his team were interested in unravelling behavioural patterns that characterise a 'good pair', with the aim of using behaviour as a predictor of a pair's reproductive success (see also Chapter 4).

2.3.1 Relationship structure of the Seewiesen flock of greylag geese

In this early study, three bonding types were identified: individuals without a primary pair partner ('unpaired'); 'monogamous' – individuals with one long-term pair partner; and individuals in a 'trio', which were bonded with two long-term partners throughout the year. Among the trios no additional liaisons were observed, while secondary liaisons were relatively frequent among unpaired geese and those with a monogamous partner.

The most frequent types of relationship were monogamous pairs without secondary liaisons, and monogamous pairs with secondary long-term liaisons (Fig. 2.2). Most of the females that engaged in long-term liaisons had no pair bond with a primary partner and hence were essentially 'unpaired' (32 of 44 long-term liaisons, 73%), whereas 58% of the males in long-term liaisons (28 out of 48 males) were primarily engaged in a monogamous pair bond (Fig. 2.2). Only 2% of all observed bonding types were trios which were stable over the year. In general the 'typical' female options were either to become a primary partner in a monogamous pair and not to engage in a liaison (31% of all female relationships), or to become the secondary partner in a long-term liaison of a male in a primary pair bond with another female and not to engage in a primary pair bond (28%). However, this was not an exclusive pattern, because a few females from monogamous pairs living in a primary bond also engaged in either short-term or long-term secondary liaisons. The 'typical' male option was monogamous in a primary pair bond without liaison (25% of all male relationships) or monogamous and repeatedly engaged in a secondary long-term liaison (also 25%; Monogamous LL in Fig. 2.2).

2.3.2 'Successful' and efficient relationships

The Seewiesen research team also attempted to compare the success of the different bonding types. Therefore, they compared three measurable variables as potential covariates of 'relationship efficiency', a mixture of social parameters and one fitness component: pair bond duration, breeding success and rank order.

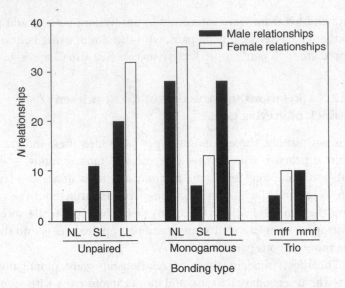

Figure 2.2   Number of observed relationships from a total of 228
relationship records based on 61 males and 71 females in the
Seewiesen flock of greylag geese between 1954 and 1969. *Unpaired*,
Monogamous and *Trio* are the bonding type categories; *NL*: no secondary
liaison, *SL*: short-term liaisons, *LL*: long-term liaisons; *mff*: one male
with two females, *mmf*: two males with one female. Note that these
are *N* cases not *N* individuals or pairs. Thus, an unknown proportion
of individuals may have contributed more than once to the emerging
pattern.

Bonds in monogamous pairs without secondary liaisons lasted
the longest: on average 5 years, with the longest relationship lasting
for 12 years (Table 2.1). When the relationships involved a secondary
liaison, pair bond durations were generally shorter. Unsurprisingly,
unpaired geese which engaged in short liaisons were at the bottom
of the scale of pair bond duration (2 years on average). Trios with two
males and one female on average remained bonded as long as monogamous pairs with a long-term liaison (4 years on average, 9 years maximum), whereas trio formations with one male and two females were
less durable (3 years on average, 5 years maximum).

A fundamental criterion for an 'efficient relationship' was
breeding success. Strictly monogamous pairs (with no engagement
in liaisons) were most successful, producing on average three offspring per year (Table 2.1, Fig. 2.3). Engaging in secondary liaisons was
related to lower breeding success: monogamous pairs with long-term

Table 2.1 *Scale of efficiency as merged from ranking pair bond duration, breeding success and social rank of pairs assigned to the different pair bond types.*

Bonding types	Efficiency	Breeding success	Rank order	Pair bond duration Mean	Pair bond duration Maximal
Monogamous pairs, no liaisons	100	3.0	High	5.4	12
Trio (mmf)	61	1.8	Medium	4.2	9
Monogamous pairs, long-term liaison		1.6	High	4.4	9
Monogamous pairs, short liaisons	40	1.2	Medium	3.5	7
Trio (mff)	21	0.4	Medium	3.0	5
Unpaired, long-term liaison		0.3	Low	3.2	7
Unpaired, short liaisons	0	0.0	Low	2.2	6
Unpaired, no liaisons		0.0	Low	2.2	4

liaisons produced 1.6 offspring per year, and monogamous pairs with short-term liaisons only 1.2 offspring per year. Trios with two males and one female (mmf) had 1.8 offspring per year, whereas trios consisting of one male with two females (mff) were less successful (0.4 offspring per year). From a singleton's point of view (i.e. for unpaired geese engaging in liaisons), repeated (long-term) liaisons with the same partner were advantageous over short liaisons (0.3 versus no offspring per year, respectively; Table 2.1).

The third criterion for breeding efficiency was position in rank order in the social flock hierarchy. Dominance ranks were assessed from monitoring the frequencies of initiating an attack and being the target of attacks. This analysis was based on the individual rank assessed for males, which was classified as high, intermediate or low. The male's rank was used as an estimate of the pair's rank because generally the female's rank was similar to, or somewhat lower than, the rank of her partner (this concept was also applied later; see e.g. Lamprecht 1991; Stahl *et al.* 2001). The researchers suggested summarising the three measures as 'breeding efficiency', because rank order co-varied in an almost linear fashion with breeding success and pair

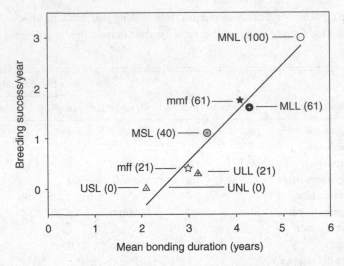

Figure 2.3   The relationship between pair bond duration and breeding success in different bonding types. It is not clear whether 'success' means hatched or fledged young (see main text). Unpaired (*U*; *triangles*), Monogamous (*M*; *circles*), Trio (*mff/mmf*; *stars*) are the pair bonding types; *NL*: no secondary liaison, *SL*: short-term liaisons, *LL*: repeated long-term liaisons; *mff*: trio of one male with two females; *mmf*: trio of two males with one female. *Numbers in parentheses* are percentage breeding efficiency (monogamous pairs without secondary liaisons taken as 100%, see Table 2.1).

bond duration. Combined, these parameters were scaled as a proportion of the values for strictly monogamous pairs (100%). Referring to this monogamous 'gold standard' for geese may be symptomatic of the moral attitudes of that time (Table 2.1).

2.3.3 Behaviour among pair partners

Heidi Buhrow and Helga Fischer were able to recognise some individuals of the Seewiesen goose flock based on specific behavioural patterns, i.e. body postures, movements or vocalisations. These researchers were extremely familiar with the individual birds after 15 years of regular, more or less daily, observations. Based on this evident individuality in behaviour they asked whether individuals who engaged in the different bonding types also differed in social behaviour (in addition to social rank; something we later called 'personality'). In other words, they asked whether the intensity of social interactions and behaviour might serve as a predictor of individual/pair breeding success. They scored the intensity of individual behaviour as 'tenuous', 'moderate' or 'intense' in each of four functional /social contexts: i.e. affiliative, sexual, parental and agonistic. This rough assessment of individual behavioural tendencies was intended as a first measure of consensus ('harmony') between pair partners and as an estimate of their behavioural matching.

In strictly monogamous pairs (in which neither male nor female had a secondary liaison) both partners had moderate expressions of behaviours in all contexts and thus were behaviourally matched to a high degree and were also more efficient in the sense that they did not show extreme behavioural intensities. Moreover, mutually responsive pair partners were more successful breeders. If only one of the pair partners scored as behaviourally 'intense' in any context, this seemed to predict additional engagement with individuals other than the primary partner. Hence, intense behaviour of a pair-bonded individual is likely to be addressed not to the pair partner but to someone else. Similarly, if a goose was scored as 'tenuous' in any one context, it was likely that their partner was directing behaviour towards more than the primary partner. At the time these data were taken, it was concluded, that behavioural moderateness and 'harmony' with the pair partner might be the essence of biological success. Today we might be critical of such results and somewhat more wary of anthropomorphic explanations.

2.4 TODAY'S PERSPECTIVE

From today's perspective, the results of these analyses are still strikingly relevant, despite some evident methodological flaws. First of all, the frequencies of the different bonding types observed at Seewiesen are also representative for the KLF flock, although today we would classify pairs with a long-term liaison partner as trios. Even more exciting is that the behavioural scores were indeed correlates of bonding type and, thus, breeding success. However, whether behaviour indeed 'predicts' reproductive success remains unclear from these correlative data, because behavioural expressions in a pair could cause reproductive success, or conversely, the reproductive potential of a pair could affect behavioural expression (Hirschenhauser 2012). Current work on pair bonding in greylag geese of the KLF flock is still following up these historical studies. Recent results support the idea that behaviour, and particularly co-variation of hormonal parameters in a pair bond, does indeed predict reproductive success (see Chapter 4). The behavioural mechanisms of pair bond formation and maintenance as well as the role of behavioural and hormonal partner compatibility are not an exclusive attribute of geese. Such data are probably relevant for monogamous species in general. In addition to genetic compatibility (Neff & Pitcher 2005), any species with social monogamy has almost certainly evolved behavioural and physiological phenomena, in particular to adapt to the biparental care of offspring and for the mutual social support of pair partners (Chapters 8 and 9). Division of labour during biparental care holds the potential for conflicts between the pair partners, and selection will favour pairs with coordinated parental efforts. The observed behaviours and hormonal patterns are probably part of the overall response to selection (Barta *et al.* 2002). In light of the sexual conflicts between pair partners, monogamy must be viewed as a compromise between conflicting evolutionary pressures on the sexes (Arnqvist & Rowe 2005).

Remarkably, 50 years ago Lorenz and his colleagues must have had a (possibly implicit) concept of the sum of behavioural characteristics of individuality. Lorenz' approach is similar to what is termed 'personality' or 'behavioural syndromes' in contemporary ethology (i.e. correlated behaviours across contexts and situations; Bergmüller 2010; Chapter 3). As in this historical example, today we also need to consider 'suites of behaviour' rather than a single characteristic for a relevant measure of individual differences. What Lorenz and his colleagues saw was real, and greylag geese still contribute to this line of research (Kralj-Fišer *et al.* 2007; Chapter 3). More than ever, the idea of consistent individual differences, for instance in fearfulness or aggressiveness, their heritability and their proximate underlying mechanisms, is a matter of ongoing debate. We conclude that goose relationships are dynamic and that the bonding types are linked with a pair's/individual's reproductive success.

2.4.1 Issues a contemporary reviewer might have commented on

In 1969 the analysis of the Seewiesen group was methodologically far ahead of its time. The data are an extraordinary and valuable legacy; a document of the era of the founders of ethology, when it was increasingly appreciated that pure observation is not enough and that quantitative data are needed. Why this work was not published in a timely manner is another question. Still, it seems highly worthwhile to dive into these data. We are honoured to present them it in this book and are grateful to Helga Fischer and the late Heidi Buhrow for contributing their classical data.

However, if this was a manuscript submitted for publication to a modern scientific journal, we – as referees – could very probably not have recommended publishing these data for the reasons addressed below. Of course, this is 'presentistic', meaning that in science and elsewhere it is inappropriate to judge historical performance from today's perspectives. However, in our case it is justified, because the methodological 'flaws' – from our present point of view – also highlight the progress made in scientific practice. Learning from such early studies and avoiding these flaws has been part of the advances of the study of behaviour. Despite this presentistic critique, we are confident that the basic patterns still hold true.

Specific remarks

With respect to the pair bond structure of the Seewiesen goose flock, for example, no details were provided on either flock size at the time of data collection or on flock management and sex ratio in the flock. The latter is particularly relevant for discussing the affiliation of secondary partners and the formation of trios. 'Pair bonding type' could refer to N individuals but at other times it seems to be about N pairs. However, assuming that there were 228 (different) pairs in the data set is also not entirely correct; instead these are 228 'relationships' based on 61 males and 71 females. This is a statistical problem known as pseudoreplication (Machlis et al. 1985). Nowadays, such a problem would be solved with multivariate statistics by including a code for each individual into the calculations. However, in the presented data set we have no information about how many individuals contributed to the data more than once, which precluded an analysis using contemporary statistical techniques.

In 1969, and particularly for Konrad Lorenz and his colleagues, it had not yet become standard scientific practice to use statistical tests to accept or reject a hypothesis. The data, as they are presented here, would be the first part of the required exploration of data distribution and homogeneity of variance as preconditions of statistical analysis. Today, the results would need to be presented with statistical measures of effect size and significance level. Nowadays we would also ask for a measure of variation and/or the range whenever an average value is presented.

Furthermore, it remains unclear how breeding efficiency was determined and why monogamous relationships with no secondary liaisons were taken as the 'gold standard' for geese. At the time of Konrad Lorenz, marital fidelity was a desirable moral value; hence the influence of ideology on this choice cannot be ruled out. In relationships with secondary liaisons, pair bond durations were generally below this monogamous 'gold standard' (Table 2.1). It remains unclear whether it made a difference which of the pair partners had the liaison. Was faithfulness more or less relevant on the female or the male pair partner's side, or both equally? Did previous breeding success influence the likelihood of engaging in extrapair (secondary) liaisons?

It is not explained whether 'breeding success' is based on the number of goslings hatched from the eggs, or on the number of goslings that successfully reached the age of fledging. This differentiation could make a great difference because many pairs have high hatching rates but fewer have successfully fledged offspring (Hemetsberger 2002; Chapter 1). Furthermore, we assume that breeding success rates for the bonding types with temporary secondary liaisons (see Table 2.1 and Fig. 2.3) refer to the primary pair partners. The comparison of breeding success between pairs with or without secondary liaisons should include breeding success with secondary partners and, strictly speaking, should be disentangled for male and female pair partners.

2.4.2 Paving the way for a new era

The knowledge of Heidi Buhrow, Helga Fischer and Jürg Lamprecht was extremely helpful in starting the 'second' Grünau period in 1990, after Konrad Lorenz had passed away in 1989. Discussions with our colleagues at Seewiesen helped in handling the flock and influenced our scientific thinking and approaches. A focus was to gather detailed information on individual life histories, such as 'Who was paired with whom, and when?' or 'Who is related to whom over how many generations?'. Only because of this detailed information is it now possible to detect and predict what the *Social Life of Geese* is all about. In the chapters that follow we try to answer various questions, many of which had already been raised by Konrad Lorenz (1991).

SUMMARY

The origins of ethology and the profile of Konrad Lorenz as a researcher are closely connected with studying greylag goose behaviour. In this chapter we present an original, previously unpublished, data set from Konrad Lorenz and co-workers on pair bond types in the greylag goose flock at Seewiesen. This also exemplifies the methodological changes in science since the early days. Descriptive and semi-quantitative observations led to measuring, and thus quantifying, behaviour and using modern statistics. Heidi Buhrow and Helga Fischer described the social structure of different pair bond types in the Seewiesen flock between 1954 and 1969. They defined primary pair partners, as well as paired or unpaired geese with temporary liaisons with secondary partners. In an attempt to rank the most successful and efficient pair bond type, they merged data on pair bond duration, breeding success and rank order. They found that strictly monogamous pairs with no engagement in liaisons were the most successful, while engagement in secondary liaisons co-varied with lower success rates. However, long-term liaisons were advantageous over short-term liaisons in terms of social rank and breeding success. Finally, scoring the intensity of typical individual behaviours led them to conclude that similar behavioural styles of pair partners co-varied with breeding success. Although we point out a number of problematic aspects in these data, we appreciate that some of the questions (such as 'What are the behavioural mechanisms of pair bonding?', 'Does pair bond type or pair bond duration co-vary with the pair's success?' and 'How do we express consistent individual differences in behaviour?') are still matters of current debate.

Part II From individual to clan

SIMONA KRALJ-FIŠER, JONATHAN NIALL DAISLEY
AND KURT KOTRSCHAL

3

Individuals matter: personality

Humans show consistent individual differences in personality (John *et al.* 2008). For example, some people are more outgoing, while others are more reserved. Recently, similar variations in behavioural tendencies have been documented in a number of animal species including molluscs, arthropods, fish, lizards, amphibians, birds and mammals (Bell *et al.* 2009). Such variation in behavioural responses to the same environmental stimuli has been referred to variously as 'coping style' (Koolhaas *et al.* 1999), 'temperament' (Réale *et al.* 2007), 'behavioural syndrome' (Sih *et al.* 2004) or 'personality' (Gosling 2001; Dingemanse & Réale 2005). All these terms imply between-individual differences and some degree of individual consistency in behaviour and physiology over time and in different contexts. Individual differences that are related to personality consist of more than one set of correlated characteristics; they can show themselves either as different behaviours or as one type of behaviour expressed in different situations (Groothuis & Carere 2005). In this chapter, we use the term 'personality traits' for the behaviours of an individual that are repeatable and correlated across context, and 'personality' for a suite of such personality traits. We also use the currently utilised term 'behavioural syndrome', which is defined as a suite of correlated behaviours that reflect intra-individual consistency in behaviours across two or more situations (Sih *et al.* 2004). While personality type is an individual characteristic, the term behavioural syndrome describes a characteristic of a population (Sih *et al.* 2004). Personality traits are moderately heritable (van Oers *et al.* 2005a),

The Social Life of Greylag Geese: Patterns, Mechanisms and Evolutionary Function in an Avian Model System, ed. I. B. R. Scheiber *et al.* Published by Cambridge University Press. © Cambridge University Press 2013.

are often contingent on physiological and neuroendocrine processes (Koolhaas *et al.* 1999; Carere *et al.* 2003; Øverli *et al.* 2007; Kralj-Fišer *et al.* 2010b), and are relevant to fitness (Dingemanse & Réale 2005; Smith & Blumstein 2008).

By the middle of the last century, Konrad Lorenz was already claiming that greylag geese differed in their behavioural propensities and that each goose had its own personality (Lorenz 1991). In the last 10 years these claims have been given further empirical support following our investigations into personalities in the KLF goose flock, from the phenomenological as well as the developmental and physiological points of view (Daisley & Kotrschal 2000; Daisley *et al.* 2005; Kralj-Fišer *et al.* 2007, 2010b; Wascher *et al.* 2008b). The goose flock at the KLF has proved to be an optimal research model for investigating personality expression, because these geese can be approached easily to enable behavioural observations, non-invasive physiological measurements (hormonal metabolites in droppings) and for experimental manipulations to be carried out (Chapter 1). Data collected on these free-ranging geese may also provide results that are more pertinent to the 'natural' situation than data collected from animals reared and tested in the laboratory.

3.1 (IN)-CONSISTENCIES IN WITHIN- AND BETWEEN-INDIVIDUAL BEHAVIOURAL VARIATION

The contextual generality of behavioural variation is a factor commonly used in the definition of personality. Contextual generality refers to the extent to which behavioural scores exhibited in one context are correlated with behavioural scores expressed in one or more different contexts for the same individual (Sih *et al.* 2004). It applies both to within- and across-personality traits. For example, aggressiveness scores in a mating context may correlate with aggressiveness scores in a foraging context (Johnson & Sih 2007). Similarly, it implies correlations between behavioural traits, e.g. between aggressiveness in one context and boldness in another context (Johnson & Sih 2007). Temporal consistency is another common characteristic of personality. Here we refer to 'differential consistency', i.e. the extent to which differences across individuals in a certain behaviour measured in one context are maintained over time. This is usually estimated using repeatability tests (Lessells & Boag 1987).

The above criteria for personality are often not strictly fulfilled, since temporal stability over the short term does not preclude personality changes over a longer time period (Stamps & Groothuis 2010a). In fact, recent studies have shown that temporal consistency of personality traits tends to decline as a function of inter-test interval duration (Stamps & Groothuis 2010b). Personality traits may change over time due to experience (learning), environmental and seasonal variation, and early social environment (Carere *et al.* 2003; Dingemanse *et al.* 2009; Naguib *et al.* 2010; Stamps & Groothuis 2010a, 2010b). Several factors may change personality during development, in particular during early development. Furthermore, contextual generality is often not observed; this is because the expression of a personality trait often depends on the context in which it is measured (see work on pumpkinseed sunfish, *Lepomis gibbosus*: Coleman & Wilson 1998; great tits, *Parus major*: see below). Behavioural traits, or the relationships between different traits, commonly differ when an individual is tested alone compared with when it is tested in a group context (e.g. pigs, *Sus scrofa domesticus*: Jensen 1995; bighorn sheep, *Ovis canadensis*: Réale *et al.* 2000; great tits: van Oers *et al.* 2005b) pointing to the fact that social mechanisms are capable of influencing both behaviour and physiology, and are likely to change the expression of personality traits (Mendl & Deag 1995).

In addition, behavioural consistency may be somewhat problematic in defining personality traits because some individuals are more stable in their behaviours than others (Bem & Allen 1974; Kagan *et al.* 1988). In particular, there may be more consistency towards the extremes of personality traits, with more variability for those individuals that tend towards the middle of a 'trait value'. Behavioural and physiological consistency has indeed most credibly been shown in animals that had been genetically selected for extreme phenotypes. In pigs, for example, only extreme phenotypes exhibited behaviours that fit the criteria of personality definition (Ruis *et al.* 2000). The coping style theory suggests that individuals differ in their behavioural flexibility, in which aggressive individuals readily form routines, while non-aggressive individuals exhibit behavioural flexibility (Koolhaas *et al.* 1999). Thus, non-aggressive individuals notice and respond to small differences in their environment, whereas aggressive individuals follow their behavioural routine.

In one of our studies we searched for short-term contextual and temporal consistency of behavioural and physiological patterns in adult male greylag geese living in an intact social environment. To explore how different test situations, social embedding and seasons influenced an individual's behaviour and physiology, we tested ten focal animals in three different seasonal phases. Geese were tested in an individual context with a handling test for the characterisation of boldness (Carere & van Oers 2004; Uitdehaag *et al.* 2008; Fucikova *et al.* 2009) and when within their social environment in different flock situations:

1. a low-density feeding situation
2. a high-density feeding situation
3. a low-density post-feeding situation
4. during rest.

Greylag males exhibited different behavioural scores between different test situations and different testosterone concentrations between seasons (Kralj-Fišer *et al.* 2007). Nevertheless, between-individual variation remained unchanged (Kralj-Fišer *et al.* 2007). Greylags showed consistent individual differences in a range of behavioural characteristics, including boldness, aggressiveness, alertness, attachment to mate and sociability (Kralj-Fišer *et al.* 2007). Among these, aggressiveness was the most stable personality trait being repeatable over time and consistent across different social situations (Kralj-Fišer *et al.* 2007). This suggests that aggressiveness may be a heritable personality trait in greylag geese, as it is for example in the house mouse (*Mus musculus domesticus*: Koolhaas *et al.* 1999), great tits (van Oers *et al.* 2004), vervet monkeys (*Chlorocebus pygerythrus*: Fairbanks *et al.* 2004) and many others (Bell *et al.* 2009). In a subsequent study, Weiß and co-workers indeed found that variation in greylag goose aggression rates could be attributed, among other things, to a heritable component and consistent inter-individual differences (see Chapter 7; Weiß & Foerster 2013). Despite a generally high level of stability in behaviour across the birds, we found higher levels of stability in male geese with extreme behavioural phenotypes. Ganders with intermediate levels of aggressiveness varied significantly more in their aggressive behaviour than the two most and the two least aggressive individuals (Kralj-Fišer *et al.* 2007), which is in line with other studies mentioned above (Ruis *et al.* 2000; Carere *et al.* 2005). Highly social individuals, however, showed very flexible behaviour, as exhibited by long and diverse behavioural sequences. These were analysed using the software package THEME (Magnusson 2000). THEME searches for repeatable behavioural patterns (i.e. routines) within an individual's behavioural flow.

Our experimental protocol, where geese were observed both when isolated from the flock and when embedded in their social environment, allowed us to test to what extent the behavioural responses of individuals are influenced by social relationships (von Holst 1998; Scheiber *et al.* 2005a; 2009a) and, conversely, how much personality characteristics affect social relationships (Hinde 1979). Assuming that active coping with a challenge in social and non-social conditions may be related to the same aspect of personality (Hessing & Hagelso 1993; Hessing *et al.* 1994; Koolhaas *et al.* 1999, 2001), we were interested to see whether social context would change the expression of this trait. To test this, resistance levels during handling were correlated with levels of aggression exhibited in the flock situations (see items 2–4 in the above list; e.g. Hessing & Hagelso 1993; Hessing *et al.* 1994). Indeed, aggressiveness within the flock tended to correlate with boldness when geese were handled individually (Kralj-Fišer *et al.* 2007), implying that the aggressiveness of the geese was somewhat modulated, but not overridden, by the social environment (Kralj-Fišer *et al.* 2007; see also Chapter 7).

3.2 PERSONALITY AS A SUITE OF CORRELATED BEHAVIOURS

Personality traits come in packages. Personality comprises suites of correlated behavioural traits, which are referred to as 'behavioural syndromes' (Sih *et al.* 2004). These vary in complexity from correlations of as few as two behaviours (e.g. aggression and boldness) to personality dimensions of several behaviours (Gosling & John 1999). Some personality dimensions may be species-specific, whereas most explorative tendencies of the most fundamental traits – e.g. aggressiveness, boldness, activity and sociability – are universal across species (Gosling 2001). In species selected for extreme behavioural phenotypes, individuals may be categorised by their behavioural and physiological effort to cope with mildly stressful situations (Koolhaas *et al.* 1999); these individuals have been found to vary along a 'reactive–proactive' dimension (wild house mice: Benus *et al.* 1991; domestic pigs: Hessing & Hagelso 1993; great tits: Groothuis & Carere 2005), comprising correlated behaviours such as aggression, activity, fearfulness, exploratory behaviour, response to environmental change, and social attachment.

Behavioural correlations are usually analysed using multivariate analysis, such as factor analysis (Bell 2007). This aims to find out which of the many variables are correlated. Variables that are correlated are loaded on the same factor. Factor analysis performed on behavioural scores in greylags suggested two factors that reflect the 'aggressiveness' and 'sociability' personality dimensions. Specifically, aggression correlated with other personality characteristics; the more aggressive males were less often attacked and showed fewer retreats in agonistic encounters, were bolder, more alert, and spent more time in close proximity to their mates (Kralj-Fišer et al. 2010b). Furthermore, aggressiveness correlated positively with dominance rank (Kralj-Fišer et al. 2010b). Sociability was a personality characteristic of its own; geese were (non-)sociable regardless of their level of aggressiveness (Kralj-Fišer et al. 2010b).

3.3 PHYSIOLOGICAL PROCESSES UNDERLYING VARIATION IN PERSONALITY TYPES

A proximate mechanism for the existence of individual variation in suites of behaviours is termed *pleiotropy* (Sih et al. 2004), where a correlation between behavioural traits is probably the result of common regulatory processes, common genes and/or common physiological processes that regulate different behaviours. Personality differences in several species were shown to be related to differential modulation of stress systems: direct or indirect measures of catecholamine (Korte et al. 1997; Sgoifo et al. 2005) and corticosterone levels (Koolhaas et al. 1999; Carere et al. 2003; Øverli et al. 2007; Baugh et al. 2012) as well as to variations in androgen levels (Koolhaas et al. 1999; van Oers et al. 2011). For instance, aggressive mice have high catecholamine and testosterone (re)activity and low corticosterone (re)activity, while the opposite is true for non-aggressive mice (Koolhaas et al. 1999). However, in contrast to the above laboratory studies, research in unselected animal lines or semi-free animals failed to find a significant correlation between personality traits and HPA axis (re)activity (e.g. wild-type rats, *Rattus norvegicus*: Sgoifo et al. 1996; shelter dogs, *Canis lupus familiaris*: De Palma et al. 2005; European rabbits, *Oryctolagus cuniculus*, in a semi-natural environment: Rödel et al. 2006; but see Cockrem et al. 2009). We found the greylag system particularly interesting because social embedding may modulate neuroendocrinological processes, including the stress response (Scheiber et al. 2005a, 2009a; Kralj-Fišer et al. 2007; Wascher et al. 2008b) and the question was whether we would still find manifestations of personality on top of that.

In the first instance, we were interested in whether greylag geese exhibit physiological and neuroendocrine consistency in processes which govern the expression of personality traits, for instance excreted immuno-reactive corticosterone metabolites (CORT, sometimes also called BM, see Kralj-Fišer et al. 2007, or CM, see Hirschenhauser 1998), which was shown in Adelie penguins (*Pygoscelis adeliae*: Cockrem *et al.* 2009). In fact, greylag geese exhibit consistent individual differences in baseline and stress-induced concentrations of CORT, consistent individual differences in heart rate (HR), and some individuality in faecal immuno-reactive testosterone metabolite concentrations (TM; Kralj-Fišer *et al.* 2007). Taking into account that the stress response of geese and their testosterone concentrations are repeatable and of limited plasticity under the same conditions, they are probably heritable and partly control and contribute to the maintenance of personality in greylags.

Assuming that pleiotropic effects of the same physiological processes may have behavioural consequences, we correlated physiological and behavioural values. We indeed found such correlations between personality traits and physiological processes in greylags (Kralj-Fišer *et al.* 2010b). Notably, the aggressiveness dimension was correlated with the stress-induced CORT concentrations to a considerable degree (Kralj-Fišer *et al.* 2010b). This implies that aggression, boldness, escape behaviour, alertness and attachment to one's mate probably all have some common proximate foundation in CORT modulation.

Interestingly, the positive correlation between the aggressiveness dimension and stress-induced CORT concentrations contrasts with the findings from other personality studies (Koolhaas *et al.* 1999; Carere *et al.* 2003; Cockrem *et al.* 2009), where the more aggressive individuals exhibit lower CORT reactivity than the less aggressive ones. Similar to these studies, baseline CORT concentrations negatively correlated with a male's aggression (Kralj-Fišer *et al.* 2010b). This discrepancy may be explained by differences in the experimental manipulation; that is, how an individual perceives the experimental situation (Sapolsky 1997). Based on measured CORT concentrations, a regular situation seemed to be more stressful for non-aggressive (low-ranking) ganders, whereas certain challenges caused higher corticosterone responses in aggressive (high-ranking) males. For example, during the handling test, when a male was removed temporarily from its social environment (the flock), the more aggressive (dominant) males produced a higher CORT response. Similarly, aggressive (high-ranking) males had higher CORT responses during high-density feeding. A logical explanation is that such situations were more socially 'demanding' for aggressive individuals, perhaps because they were unable to control local social interactions during high-density feeding to the same extent that they are able to in less dense feeding situations. Thus, our results suggest that a correlation between stress hormones and personality type is affected by the experimental (and social) context (i.e. how an individual perceives a situation; see Chapter 8), which might be stressful for one but not another personality type (Kralj-Fišer *et al.* 2010b). This is in alignment with the fact that stress responses are always very context-dependent (Kotrschal *et al.* 1998a).

In a range of primate species it was found that stress level and social rank were related as well. However, in this instance and depending on the circumstances, individuals with the highest and lowest ranks were generally most stressed (Abbott *et al.* 2003; Sapolsky 2005). Another possible reason for a higher CORT response in the more aggressive ganders might relate to temporal factors; it is likely that short-term CORT elevation promotes and primes aggressive and active responses in the near future, whereas chronic elevation inhibits aggression and promotes subordinate behaviour (Summers *et al.* 2005). In other words, steady subordination might cause ineffective CORT responses to stressful situations. This should be explored further.

Individual greylag geese also differed consistently in their heart rate (HR) parameters, but we found no significant correlation between mean heart rate and personality (Wascher *et al.* 2008b; Kralj-Fišer *et al.* 2010b). However, a more detailed analysis revealed that the more aggressive, bold, vigilant and mate-attached the males were, the greater the HR increase during aggressive interactions (Kralj-Fišer *et al.* 2010b), which is in agreement with the proactive–reactive continuum (Koolhaas *et al.* 1999). Thus the modulation of HR is very likely to be another underlying mechanism of aggressiveness in geese, although this relationship needs to be investigated further. Another personality axis, sociability, was related to testosterone modulation. Low testosterone levels were correlated with high sociability within the flock. This suggests that high testosterone negatively interferes with flock integration.

3.4 PERSONALITY AND REPRODUCTIVE SUCCESS: PRELIMINARY DATA

Variation in personality is often related to differences in fitness in natural populations (Dingemanse & Réale 2005; Smith & Blumstein 2008 for reviews). We attempted to examine this question using ten focal individuals. Although this number is quite low for reliable predictions, such preliminary data might give us some ideas for future studies. The other shortcoming is that these results are indirect, since both parents may influence their offspring's behaviour. However, in geese, the male of a pair may directly (via sperm and/or genetic quality) or indirectly (via social support) have an influence on the reproductive success of the pair (Lamprecht 1986b). During the breeding season, one role of the gander is to guard his female partner during egg-laying and incubation. After the young have hatched, the gander also plays a major role in guarding the goslings (Lamprecht 1986b; Lorenz 1991). A gander may also influence the clutch size prior to the mating season by allowing the female partner to concentrate on feeding (Lamprecht 1986b) and by providing social support, which modulates stress hormones and affects energy expenditure (Scheiber *et al.* 2005a; 2009a). Despite our small sample size, we indeed found a relationship between the ganders' affiliative tendencies and reproductive success. Pairs that were more attached to one another hatched and fledged more offspring than loosely bonded pairs that spent more time with other flock members. This is not surprising, since an attached male spends most of his time guarding his female partner during the egg-laying period and incubation as well as providing parental care, which gives such a pair a better chance of reproductive success.

3.5 EXPERIMENTAL MANIPULATIONS: ENHANCED YOLK TESTOSTERONE AND ITS EFFECT ON BEHAVIOUR

Variation in personality traits is partly explained by heritable factors. However, in addition, as so often is the case, experience can influence personality traits at any time of life. Particularly strong influences on personality development occur during the early stages of life, sometimes even before birth or hatching, which is commonly called 'maternal effects' (Groothuis & Carere 2005). For example, in birds, females deposit steroid hormones in the egg (mostly in the yolk) during oogenesis. The amounts of the two steroids, testosterone and corticosterone, that end up in the egg appear to be influenced by the female's behavioural and physiological state at the time when the layers of fat and protein that will eventually form the embryo's nutrients are being laid down. These steroids have been shown to exert profound organisational effects on the embryo's nervous system, shaping the development of neuronal connections (Sapolsky 1992; Fuxe *et al.* 1996), thus influencing the expression of behaviours post-hatch as well; chicks exposed to an increased level of maternal testosterone may beg for food more often and are more aggressive in the nest (Schwabl 1993; 1996a), show increased exploratory behaviours (Gvaryahu *et al.* 1986), and show differences in learning and discriminatory behaviours (Morris et al. 1971; Clifton et al. 1988). Maternal influence over the amount of steroid deposited in the egg suggests that females may be able to manipulate certain characteristics of their offspring; for example, the effects of differential yolk testosterone allocation may be directed at other functions such as foraging efficiency and tactics (Marchetti & Drent 2000; Fritz & Kotrschal 2002; Pfeffer et al. 2002). This points to early steroid exposure being causally linked to behavioural phenotype and individual coping style.

We looked at the influence of early steroid exposure on greylag geese by examining the effects of experimentally augmenting testosterone in the yolk on the development and expression of individual behavioural phenotypes. In addition, we investigated the steroid status of individuals in relation to behavioural challenges, as determined by non-invasive faecal sampling. We expected that experimentally enhanced yolk testosterone would affect growth, phenotype and responsiveness to stress.

To determine whether there was a valid premise for our investigations, it was necessary to find out whether there were differences in the levels of steroid deposited naturally within clutches of goose eggs. Goose eggs from our own flock and from clutches from a captive flock at the University of Groningen were analysed for steroid content using an enzyme immunoassay developed for testosterone determination at the Veterinary Medical School in Vienna (Palme & Möstl 1993; Kotrschal et al. 1998a; Hirschenhauser et al. 2000). On analysis, it seemed that testosterone values did indeed vary both between clutches of eggs and also within a clutch (mean ± SEM = 4.159 ± 1.8 ng testosterone metabolite (TM) /g yolk). When corrected for within-clutch variation, it was apparent that the second-laid egg showed a tendency to have a greater concentration of testosterone than the first-laid egg, with the fourth-laid egg having less than both the first and second-laid eggs (Fig. 3.1). These findings prompted us to suggest that there was differential testosterone deposition within a clutch and that these differing amounts would possibly be sufficient to produce differences in the behaviours of the offspring.

The next stage was to apply these findings with individuals from the KLF flock, namely to test for any such behavioural/personality differences that would be concomitant with differing egg testosterone levels. For the planned work, we used hand-raised animals as our experimental and control subjects both during the hand-raising period and following their integration into the flock. Fertilised eggs were taken

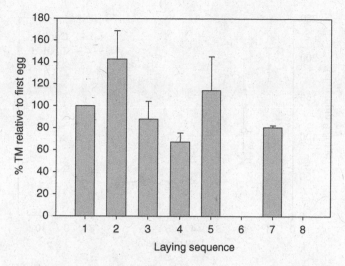

Figure 3.1   Testosterone metabolite (TM) levels (%) in the egg yolks of greylag goose eggs relative to the first-laid egg in each clutch. Number of clutches ($N = 9$) from which 1–8 eggs were taken. Values expressed as a level relative to the first-laid egg, taken as 100%, are shown for each egg within the clutch. Due to insufficient data, the levels from the sixth- and eighth-laid eggs are not shown.

from laying females of our flock. In order to augment testosterone levels, some of the eggs received an injection of an additional physiological dose of testosterone in ethanol as a solvent, which would raise the concentration of the steroid to the higher physiological levels found in the eggs we had previously analysed (Daisley & Kotrschal 2000). All the eggs were then transferred to an incubator, with the goslings being taken by their human foster parents immediately upon hatching. In total 8 (5 male, 3 female) goslings hatched successfully from treated eggs, with 19 (8 males, 11 females) serving as controls.

The steroid hormone profiles for these goslings were assessed throughout the hand-raising period until the birds fledged. This meant that developmental and baseline steroid hormone metabolism could be measured. In addition, we examined the effects of environmental 'challenges', such as the birds' responses to 'stressful' events in terms of changes in metabolism below or above each individual's background levels.

Shortly after hatching, faecal TM levels were high, but rapidly decreased thereafter (see also Frigerio *et al.* 2001b). Individuals with *in ovo* testosterone supplementation ('treated') excreted more TM

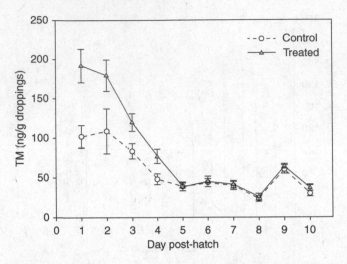

Figure 3.2   Testosterone metabolite (TM) levels determined by EIA from the droppings of testosterone-treated and control greylag geese. *Dotted line/open circles*: controls (*N* = 18); *solid line/filled triangles*: treated animals (*N* = 8). Values of ng steroid per g droppings (*y*-axis) plotted against day post-hatch. Mann–Whitney *U*: treatment effect on days 1 and 2; *p* < 0.05.

during the first 5 days following hatching than did the controls (Day 1, controls 103 ± 17 ng TM/g droppings ± SEM, treated 193 ± 22 ng/g; Mann–Whitney *U*, *z* = 3.06, *p* = 0.002; Day 2, controls 114 ± 31 ng/g, treated 178 ± 18 ng/g; Mann–Whitney *U*, *z* = 2.1, *p* = 0.033) (Fig. 3.2). Levels in both groups then declined and after the fifth day post-hatch there were no changes associated with gosling age. There were also no apparent baseline differences between the two groups after the fifth day of hatching.

The excretion of corticosterone metabolites followed a similar pattern, being very high on hatching and then declining in value to reach a more or less constant level after 5 days. No significant differences in CORT excretion were found between the treated and non-treated birds, however (Mann–Whitney *U*, *p* > 0.05). Basal excretion of TM and of CORT was again analysed at 6 weeks of age. There was no significant effect of treatment for either of the steroid metabolites (controls 15.2 ± 2.5 ng CORT/g droppings ± SEM, treated 11.6 ± 3.5 ng/g; $F_{1,16}$ = 4.16, *p* = 0.083; controls 688 ± 107 pg TM/g droppings, treated 876 ± 216 pg/g; $F_{1,25}$ = 0.679, *p* = 0.497), suggesting that the underlying metabolic rates and production of steroid were similar between

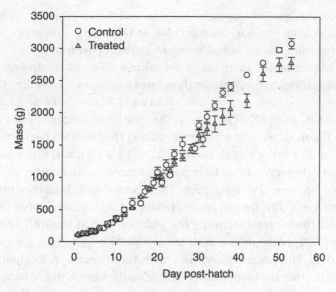

Figure 3.3   The mass of goslings (g) as recorded from day 1 post-hatch until day 53 (close to fledging). *Open circles*: controls (N = 14); *filled triangles*: treated goslings (N = 8).

treated and untreated birds. These initial differences in steroid metabolite levels were also reflected in the growth and physical characteristics of the goslings; towards the later stages of hand-raising there was a tendency for the control goslings to be heavier shortly before fledging, i.e. between days 50 and 60, compared with the treated individuals (treated 2,491 ± 307 g, controls 2,878 ± 184 g; Mann–Whitney U, *p* = 0.086; Fig. 3.3). At all other times we found no differences.

To investigate the effects of additional *in ovo* testosterone on the goslings, we carried out a range of simple behavioural tests. We collected samples of droppings before and after the tests. As part of the suite of tests, we instigated an 'open-field' individual isolation. Open-field tests are regularly used to assess individual stress responses and locomotor activity both in laboratory animals and in field studies (Verbeek *et al.* 1994). Here we isolated goslings, individually, from their foster parent and other family members by taking them to a rectangular box positioned away from auditory and visual contact. The gosling was then left in the box for a short period (5 min) after which it was returned to the family group. During its isolation, the goslings' behaviours were recorded remotely by a video camera. The goslings were assessed for their latency to start distress-calling, the number of calls emitted, the level of locomotor activity, and the number of attempts to escape from the box. On returning to the group, droppings were collected from the individual for up to 120 min post-test.

Behaviourally, there was a significant effect of treatment; the treated goslings had a significantly greater latency to start calling than did the control birds (treated 1.56 ± 0.44 s $\pm$ SEM, controls 0.88 ± 0.06 s; $F_{1,24} = 5.546$, $p = 0.027$). Conversely, the treated goslings showed a tendency for an increase in locomotion during the isolation trial (treated 272.96 ± 20.1 units $\pm$ SEM, controls 222.63 ± 12.9 units; $F_{1,24} = 3.214$, $p = 0.084$). However, there were no differences in the number of calls emitted between the two groups. There was also a difference in the excretion of CORT; treated goslings excreted significantly more CORT than did the controls following the isolation period (controls 18.61 ± 2.9 ng CORT/g droppings $\pm$ SEM, treated 32.72 ± 7.1 ng/g; $F_{1,25} = 7.326$, $p = 0.018$). This seems to correlate with the increase in locomotion, suggesting that the treated birds were actually more active in escaping from their holding area, although the increased latency to call and no increase in call number during the isolation would seem to suggest that the treated birds did not perceive the trial as any more stressful than did the controls.

To determine the effects of absence of the foster parent, the goslings were walked to an outdoor arena and provided with a shaded area and water and left there, away from visual contact from their human foster parent, for a 20-minute period. Droppings were again collected both before and after the separation period. We found significant differences between the treatments in CORT excretion; post-isolation there was a trend for the treated goslings to excrete lower CORT than the controls (controls 35.36 ± 3.9 ng CORT/g droppings $\pm$ SEM, treated 22.68 ± 2.3 ng/g; ANOVA, $F_{1,25} = 3,548$, $p = 0.073$). When comparing pre- and post-test measurements, the CORT levels for the control group post-test were raised compared with pre-isolation values, whereas in the treated goslings the amount of excreted CORT was reduced post-test (controls $264 \pm 167\%$, treated $88 \pm 28\%$; $F_{1,25} = 5.384$, $p = 0.03$; Fig. 3.4). This would indicate a decreased 'stress' response in the treated goslings compared with the controls.

The ability to learn requires moderate increases in corticosterone in order for memory formation to occur (Sandi & Rose 1994). We have seen that there are differences between the treated and control birds in terms of their stress response, or at least in terms of their steroid metabolism. We provided tasks aimed at testing for differences in individual learning performance. We used a methodology developed

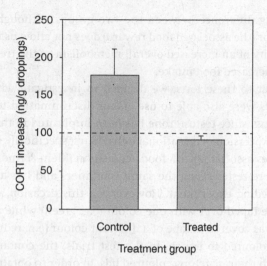

Figure 3.4   Changes in immuno-reactive corticosterone metabolite (CORT) excretion following group isolation in testosterone-treated and control goslings. Control group (N = 18); treated goslings (N = 8). Mean values of CORT (ng/g droppings, y-axis) post-isolation are divided by pre-isolation levels. *Dashed line* indicates pre-isolation baseline.

by Pfeffer *et al.* (2002) in which the ability of goslings to access food hidden in containers was determined. Goslings were pre-trained to take food from small, opaque, open-topped plastic containers. Having learned the association between container and food, the goslings were individually presented with four such containers, this time with the food hidden by a removable, opaque plastic lid. The goslings had to remove the lids in order to access the food. The birds would nuzzle, poke and bite the containers and lids, and would eventually 'shape' their behaviours in such a way that the lids could be removed. The number of attempts made before each gosling was able to remove a lid was recorded. Subsequently, the number of test trials taken for the goslings to remove the lids of all four containers within the allotted test period (30 s) was taken as a measure of their learning performance. We found that the time taken to approach the containers did not differ between the control and treated groups, which meant that any subsequent differences were not the result of fearfulness associated with approaching a novel object. Once the goslings started to manipulate the lids, differences were apparent between the groups, with the testosterone-treated birds needing significantly fewer trials to remove all four lids (Mann–Whitney U, $z = 2.323$, $p = 0.021$). This suggests a real learning difference between the two groups, although a motivational factor (the associated food reward) does not allow us to rule out the possibility of an increased overall metabolism of the treated birds requiring increased food intake.

Further to these tests, we decided to investigate whether the treated birds were also able to use colour discrimination to enhance their learning, since testosterone has been implicated in the development and expression of attentional behaviours, specifically in relation to visual cues associated with food acquisition (Klein & Andrew 1986). Goslings were tested using the same containers and lids as outlined in the preceding experiment. However, on this occasion, three containers were taped over with one colour (e.g. green) while the fourth container was covered in tape of a different colour (e.g. red); their lids were also coloured to match. In the test trials, the containers were covered with their matching coloured lids. In order to obtain food, the goslings had to uncover the lid of the uniquely coloured container (i.e. red); the other three containers held no food. The goslings, therefore, had to learn that the differently coloured container held the food. We recorded the time taken for the goslings to first touch the containers, the number of trials until the test had been learned, and the number of containers tried prior to successful lid removal. There were no differences between the two treatment groups in terms of time taken to approach the containers; however, the testosterone-manipulated individuals carried out the task with fewer errors (Kruskal–Wallis: $\chi^2 = 5.35$, $p = 0.021$; Fig. 3.5). In both groups, increases in CORT excretion followed the learning task (control: pre-test 3.2 ± 1.7 ng CORT/g droppings $\pm$ SEM, post-test 7.8 ± 2.9; Wilcoxon $p = 0.005$; treated: pre-test 2.9 ± 1.5, post-test 7.3 ± 2.7; Wilcoxon $p = 0.002$), although we found no differences between the groups. Thus, elevated testosterone in an embryo evidently leads to improved learning ability. Although this learning may again be related to the motivation of acquiring food, this does suggest that these treated birds are likely to be capable of exploiting their local environment more successfully than the birds that had experienced a lower level of steroid exposure in the egg.

Taken together, these tests point to the fact that testosterone injected into the yolks of freshly laid greylag goose eggs had effects on both the behavioural and physiological characteristics of the goslings. Enhanced yolk testosterone shifted individuals towards a more 'proactive' coping style (Koolhaas et al. 1999). This is in alignment with results from Japanese quail (Coturnix coturnix japonica: Daisley & Kotrschal 2000) and indicates that the natural variation found in the yolks of goose

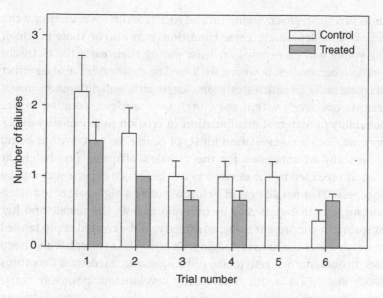

Figure 3.5   The number of green lids opened ('failures') before gaining access to a red container with food. On trial 5 of 6 the position of the red container was changed. *Open bars*: controls ($N = 14$); *filled bars*: treated animals ($N = 8$).

eggs may indeed produce significant effects on the behavioural pheno-types of hatchlings. The fact that the treated birds excreted increased amounts of testosterone metabolites during the first 5 days post-hatch compared with the controls indicates that our manipulation may have affected early set-points of steroid metabolism. Initial exposure to ster-oids as an embryo has organisational effects on the nervous system (Sapolsky 1992) and, in agreement with the idea that enhanced yolk testosterone shifts individuals towards 'proactive' coping, the treated individuals in the present study showed a reduced stress response, at least during the group isolation, in comparison with the controls.

It has been shown that the levels of steroid in the egg are under maternal control (e.g. in zebra finches, *Taeniopygia guttata castanotis*: Gil *et al.* 1999). Females adjust testosterone allocation according to both the social (Schwabl 1996b; Eising *et al.* 2001) and the physical envi-ronment (Schwabl 1996a; Gil *et al.* 1999) which they are experiencing at the time of oocyte maturation. This may predispose hatchlings to compete not only with siblings in the nest, but also within the social environment after fledging. Such maternal manipulation may provide a mechanism to facilitate rapid phenotypic adjustments of offspring traits in response to environment and social conditions operating at the time of yolk deposition. These conditions may also be those to which the young will be exposed, at least during their early life (Schwabl 1997). In our studies, however, we found no clear evidence of an effect of laying order on yolk testosterone levels, although differences in egg testosterone levels within the clutch were evident, pointing to the possibility of maternal manipulation in relation to personality development. Such a manipulation must, of course, be perceived as being biologically advantageous for the female's offspring post-hatch, at least as predicted by the environment the female experiences during oogenesis. That not all eggs in a clutch receive a high dose of testosterone suggests that costs may be inherent to both the female and her offspring. In our present study, goslings from the treated group tended to weigh less at the time of fledging; an equivalent result has also been seen in domestic chickens (*Gallus gallus domesticus*: Riedstra & Groothuis 2000). Body mass at this time during development positively correlates with survivorship in emperor geese (*Anser canagicus*: Schmutz 1993), and was also shown to be reduced in American kestrel (*Falco sparverius*) hatchlings that had developed from testosterone-enhanced eggs (Sockman & Schwabl 2000).

In conclusion, it is apparent that the KLF geese vary in their individual abilities and how they deal with environmental perturbations. There will evidently be advantages and disadvantages associated with the personality traits of the individual. Testosterone-treated individuals tend to be more proactive, with the controls possessing more reactive phenotypes. Post-fledging, boldness and aggressiveness (e.g. high testosterone) or shyness and submissiveness (e.g. low testosterone) may indirectly and negatively affect fitness, for example by interfering with the integrity of the flock or of the family unit. In addition, the economic use of resources and risk-proneness in social interactions may be affected (Silverin 1990; Drent & Marchetti 1999; Marchetti & Drent 2000). Predator avoidance may also be influenced, perhaps due to attention during foraging and higher levels of activity (Dukas & Kamil 2000) and dispersal rates. Aggression, boldness, escape behaviour, alertness and attachment appear to be related directly to CORT modulation. For example, high-ranking geese were more prone to increases in CORT during (high-density) feeding situations. We have shown that changes in egg testosterone levels can also be linked to CORT modulation. Thus, trade-offs involving testosterone (and subsequent CORT responses of the individuals) may provide additional strong selection pressures for the female to regulate egg testosterone levels. If differences in yolk testosterone correlate with behavioural phenotypes and CORT response (i.e. shifting individuals along the proactive–reactive axis), maternal manipulation of steroid levels could possibly allow a more economic exploitation of resources within the family or foraging group (Marchetti & Drent 2000).

SUMMARY

This chapter summarises various aspects of personalities (inter-individual differences in behaviour) of individual greylag geese from phenomenological as well as developmental and physiological points of view. As part of the study, we repeatedly observed ten males in various experimental situations, namely in social and individual contexts. Our observations revealed that they exhibited individual differences in aggressiveness, boldness, alertness, 'proximity to the female partner' and sociability. The first three were inter-correlated in a so-called 'aggressiveness syndrome', whereas sociability proved to be independent of aggressiveness. Perhaps this is why individual aggressiveness, which shares the same underlying trait as boldness, did not substantially change in an individual versus a social context. Further to this, we discussed physiological factors that could explain personality differences in geese. In the KLF geese, we showed for the first time that individuals differ consistently, and thus exhibit individuality in their stress response (measured as excretion of glucocorticosterone metabolites, CORT and heart rate) and also in their testosterone levels (measured as excretion of testosterone metabolites, TM). Aggression, boldness, escape behaviour, alertness and attachment were all related directly to CORT modulation, whereas TM modulation related to sociability. Hence, we suggest that CORT modulation depends on how an individual perceives a situation, rather than on personality *per se*. However, low TM levels were related to high sociability with other flock members and vice versa. Furthermore, we describe how early exposure to testosterone affects the expression of the goose's behavioural phenotype. Testosterone-treated individuals tended to be more proactive, with the controls possessing a more reactive phenotype. Post-fledging, overt boldness and aggressiveness (e.g. high TM) or shyness and submissiveness (e.g. low TM) may indirectly and negatively affect fitness, for example by interfering with the integrity of the flock. Further, we propose how changes in egg testosterone levels can also be linked to CORT modulation by means of trade-offs involving testosterone and CORT responses of the individuals. This may provide additional strong selection pressures for the female to regulate egg testosterone levels.

4

Maintenance of the monogamous pair bond

Forming and maintaining pair bonds is usually the behavioural frame which allows individuals of monogamous species to optimise resource acquisition and reproductive success. However, why do some (bird) species form long-term dyadic partnerships and others do not? Individual geese usually form and maintain a socially, and usually also sexually, exclusive relationship with another individual, which may last a lifetime. In the KLF flock, some pairs never raised offspring successfully but remained paired with each other even in the absence of reproductive success; why did these pair partners stay together as a pair? Were these geese more 'compatible' than others and how can we measure partner compatibility? The dyadic bonds between males and females may be regarded as core elements of greylag goose societies (Kotrschal *et al.* 2010). In this chapter, we focus on the heterosexual pair, although alternative tactics such as 'homosocial' male–male pairs and trios exist (Kotrschal *et al.* 2006; Chapter 5). We deal specifically with already established pairs, in particular the behavioural and hormonal mechanisms involved in pair bond maintenance and the fine-tuning between pair partners.

Following a complex and still not fully understood mate-choice process (Choudhury & Black 1993; Choudhury *et al.* 1996), a long-term bond is established between a female goose and a gander. Thus, the initial choice of a partner seems a critical precondition for a lasting relationship and for lifetime reproductive success (Dunbar & Shultz 2007). Similar hormonal and other physiological mechanisms seem to be involved in different aspects of the pair bond, such as mate attraction,

The Social Life of Greylag Geese: Patterns, Mechanisms and Evolutionary Function in an Avian Model System, ed. I. B. R. Scheiber *et al.* Published by Cambridge University Press. © Cambridge University Press 2013.

mate preference, and maintenance of the relationship in established pairs. Pair bond maintenance may be viewed as the result of 'bilateral' interactions between pair partners and of 'multilateral' interactions between the pair and their social environment. Simply focusing on individual parameters for each partner or sex separately may give an incomplete picture in studies of mating behaviour, biparental care and reproductive output, as 'it takes two to tango'. This has already been pointed out by the early Seewiesen researchers (Chapter 2). We therefore shifted the operational unit of analysis from the individual to the pair.

Continuing partnerships in geese seem to benefit from behavioural (Lorenz 1991; Lamprecht 1992; Mausz et al. 1992; Black 1996) as well as hormonal (Hirschenhauser et al. 1999b; Weiß et al. 2010b) synchrony between pair partners, for example to promote the effectiveness of biparental care. Evidence from geese indicates that this synchrony is due neither to mate choice nor to mutual adjustment over time and is affected by disturbances from other flock members (Hirschenhauser et al. 2010). The maintenance of pair bonds may also be viewed as an example of cooperation between two unrelated individuals, as both benefit from tolerating and supporting each other (Soares et al. 2010). One may think about pair partner compatibility and coordination processes between pair partners as varying degrees of cooperativeness. The variable cooperativeness of one pair partner and the potential compensation through the other pair partner may result in conflicts between the sexes over the division of labour (Balshine et al. 2002). This is particularly evident during raising of the offspring, when the coordinated interactions of the pair partners determine reproductive success (Lamprecht 1992). However, it is still puzzling why some pair partners remain paired even when they have never successfully raised offspring. As a general tenor of social relationships, including friendships and similar alliances, Cords and Aureli (2000) introduced the concept of three components of a 'qualitative relationship': its value (direct benefits, e.g. access to food), partner compatibility (the level of tolerance between individuals), and security (consistency over time). Originally based on primate studies, this concept was recently also found to be of value in a highly social bird, the common raven (Corvus corax: Fraser & Bugnyar 2010a). Our particular studies of pair partner compatibility in geese might extend the definition of a simple 'tolerance between individuals' (Cords & Aureli 2000; see also Chapter 11) and include a continuing and mutual 'active responsiveness between individuals'. When we add seasonal changes to the 'active responsiveness between partners', a complex scenario emerges with alternating behavioural roles and context-related physiological responsiveness between the individuals.

The behavioural compatibility of partners is relevant for understanding the quality of a pair bond. For example, in great tits (*Parus major*), fledgling condition is affected by both parents' exploratory behaviour, and pairs consisting of a male and a female with similar personalities had the highest long-term reproductive success (Both *et al.* 2005). In the following sections, we examine the process of pair bond maintenance as result of a continuing mutual responsiveness, which may vary both seasonally and between years. We present detailed analyses of (i) behavioural coordination and (ii) seasonal hormonal synchrony between pair partners in geese.

4.1 BEHAVIOURAL SYNCHRONY BETWEEN PAIR PARTNERS

Behavioural synchrony refers to the coordination of behaviours between individuals over time. Synchrony has been reported between individuals (Paradis *et al.* 2000) or groups within populations (Conradt & Roper 2005), and among pair partners (Canada geese, *Branta canadensis*: Raveling 1969a; blue-backed manakins, *Chiroxiphia pareola*: Snow 1977; Chinese painted quail, *Excalfactoria chinensis*: Gubler 1989; humans, *Homo sapiens*: Grammer *et al.* 1998). Over time, behavioural synchronisation processes between mates may become routines and may develop into rituals (Wilson 1975b), which probably facilitate pair bond maintenance (Wachtmeister 2001). If selection favours behavioural synchronisation between mates, evolutionary ritualisation may occur, as is the case with the courtship behaviour of some birds, such as the highly coordinated and ritualised courtship display of the great-crested grebe (*Podiceps cristatus*: Huxley 1914) or the parallel courtship flights of common ravens (Lorenz 1939b). Basically, a certain degree of behavioural coordination and synchrony may be required to keep any pair together and the relationship 'functional' over the years. It has been suggested that coordinated activities incur certain costs to the individuals involved, such as compromising one's own activity for the sake of maintaining spatial cohesion with companions (Conradt & Roper 2005). From an evolutionary perspective, these costs should be balanced against benefits such as securing a mate and jointly raising offspring (Birkhead 1980; Birkhead *et al.* 1985; Schneider & Lamprecht 1990; Davis 2002).

The time budgets of separate individuals within a group have been shown to provide a valuable insight into the level of behavioural coordination with other non-related individuals (Ruckstuhl 1999). Thus, we used time budget data to investigate the behavioural coordination between pair partners. When two individuals, in our case the pair partners, are recorded simultaneously, it can be measured how much of the time budget consists of coordinated behaviours and how these are distributed over time. Within-pair synchrony may be assessed from simultaneously performed behaviours (Spoon *et al.* 2007), but also from functionally complementary behaviours. We therefore classified behaviour patterns as follows:

- *functionally equivalent behaviour combinations*, i.e. both partners engaging in the same behaviour simultaneously
- *functionally complementary behaviour combinations*, i.e. partners engaging in different behaviours in functionally meaningful combinations, for example, the male being vigilant while the female is foraging (Gauthier & Tardif 1991; Kotrschal *et al.* 1993).

We set out to test the adaptive value of behavioural synchrony between pair partners in greylag geese. If a high degree of behavioural coordination is adaptive, it should be positively correlated with reproductive output, and would therefore differ between successful and unsuccessful breeders. In geese, a failed reproductive attempt increases the odds of reproductive failure the following year and vice versa (Lamprecht 1990; Black *et al.* 2007). This is an intriguing phenomenon in geese and may be viewed as analogous to winner/loser effects (i.e. individuals with previous experience of dominance are likely to be dominant in future encounters; Begin *et al.* 1996). Therefore, grouping pairs into 'successful' and 'unsuccessful', according to an all-or-nothing measure of reproductive success, seemed appropriate, i.e. whether or not they had raised at least one offspring to fledging in the course of time that a pair bond lasted. Current reproductive status during the year of data collection was categorised as 'parental' or 'non-parental'.

From a total of 13 focal pairs (8 successful, 5 unsuccessful), we recorded behaviour from both pair partners simultaneously during the brood care phases, and calculated the proportions of time which pair partners spent performing equivalent and functionally complementary behaviours. Throughout all the observations, the pair was regarded as a unit, characterised by a combination of the male's and female's behaviour. Over 2 years, nine pairs (eight successful, one unsuccessful) were observed with 1–3-week-old young during data collection and were labelled 'parental'; nine pairs (five successful, observed also as parental pairs in the previous year, and four unsuccessful) had no young of the year during data collection and were labelled 'non-parental'. We observed the full range of time that the pair partners were engaged in functionally equivalent behaviours from 8.5% to 98.3% of the total time of individual observations (coefficient of variation (CV) between pairs 24.5%) and also for functionally complementary behaviours (between 1.7% and 90.4%; CV between pairs 16.1%). The most frequently observed combinations were 'simultaneous vigilance' and 'simultaneous resting'. The most frequently observed functionally complementary behaviour combinations were 'male vigilant while female was resting' and 'male resting while female performed comfort behaviours'. Figure 4.1 summarises the most frequent combinations of behaviours of male and female partners throughout all recorded pairs. Overall, successful pairs spent less time engaging in the same behaviours simultaneously ($F = 7.01$, df = 1, $p = 0.02$) than unsuccessful ones and simultaneous resting bouts were shorter than in unsuccessful pairs ($F = 4.2$, df = 1, $p = 0.05$; Fig. 4.2a). While this may seem surprising at first glance, it suggests that in successful pairs individuals complement each other more than in unsuccessful pairs. Therefore this indicates a lower degree of within-pair coordination in unsuccessful breeders and probably a more efficient division of tasks in the successful ones. When mates spend more time performing equivalent activities, there is less time for complementary activities with a functional significance to survival and reproduction; for example, during simultaneous resting, the ability of both mates to engage in anti-predator vigilance and mate guarding may be impaired (Gauthier & Tardif 1991).

Successful and unsuccessful pairs generally spent similar proportions of time on a variety of functionally complementary behaviours ($F = 1.3$, df = 1, $p = 0.3$), such as 'male vigilant while female forages' ($F = 0.4$, df = 1, $p = 0.5$), a relatively common combination of functionally complementary behaviours (Gauthier & Tardif 1991). Surprisingly, successful pairs spent more time in the functional combination 'male forages while female is vigilant' ($F = 4.7$, df = 1, $p = 0.05$; Fig. 4.2b), a relatively 'unconventional' combination. It seems that pair partners who share responsibilities in an adaptive way take turns to a greater extent, in this case with the female being vigilant. This may pay off in the long run. Complementary coordination between mates is supposedly a major benefit of monogamy (Bradley et al. 1995; Cezilly &

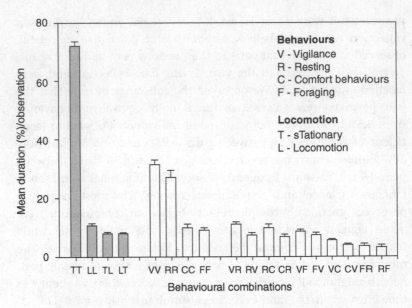

Figure 4.1   Mean proportions of total observation time (±SE) of the main combinations of behaviours. In each combination of two letters on the x axis, the letters represent the behaviour performed by the male (*first letter*) and the behaviour performed by its female mate at the same time (*second letter*). Behaviours are first grouped into actions performed during locomotion and while stationary (*filled columns*), and second into more discrete categories of behaviours (foraging, being vigilant, resting, comfort behaviours: *open columns*).

Nager 1996). In geese, an advantage of complementary coordination behaviour became apparent with long-term reproductive success (e.g. a generally successful pair or not) rather than with current reproductive output. However, dynamic sexual roles and potential sexual conflict between pair partners remain to be studied in geese.

4.1.1 Within-pair distance

As in the case of other birds, and also in primates, spatial proximity indicates social bonding in geese (Frigerio *et al.* 2001a; Chapters 6 and 11). Over the course of the entire reproductive cycle, mates may provide passive (i.e. 'emotional', in contrast to 'active', which would be interference in agonistic interactions on the side of the partner) social support to each other merely through their presence, but also by certain forms of post-conflict behaviour (Lamprecht 1986a, 1986b; Weiß

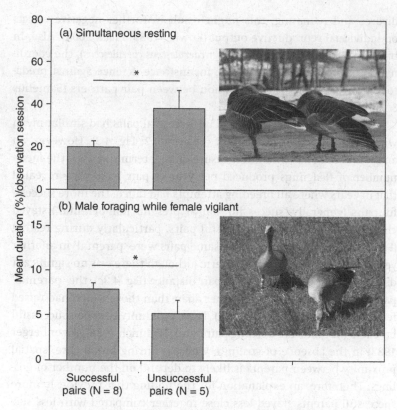

Figure 4.2   Mean durations (±SE) of (a) simultaneous resting, and (b) male foraging while female vigilant shown by the pair partners of successful and unsuccessful pairs. *Asterisks* indicate significant differences (*p* < 0.05). Photographs © I. Nedelcu.

& Kotrschal 2004; Scheiber *et al.* 2005a; Weiß *et al.* 2008; Chapter 9). In fact, spatial proximity may be the essence of social support in geese, which never allo-groom (see Chapter 11). Social support may positively influence the mate's energy budget by dampening the glucocorticoid stress response, thereby preventing metabolic overshoot due to stress (Scheiber *et al.* 2009a), and consequently may enhance future reproduction (Perfito *et al.* 2002; Angelier *et al.* 2007). Behaviour and spatial proximity to the long-term partner have a psycho-physiological effect, but also affect practical aspects, as a partner nearby may quickly provide active support in interactions if needed (Chapter 9). Loose cohesion may also facilitate pair bond challenges by rivals which, in turn, may negatively affect a pair's reproductive success (Taylor *et al.* 2001; Heg *et al.* 2003; Weiß *et al.* 2010b). Should these challenges lead to divorce and re-mating, this might result in further negative effects on individual reproductive output (Rowley 1983; Black 2001). Also, in biparental lesser snow geese (*Anser caerulescens caerulescens*), the coordination of brood care behaviours, for instance defence against predators, profited from spatial cohesion between pair partners (Samelius & Alisauskas 2001).

In our study, successful and unsuccessful pairs had similar mean within-pair distances ($F = 1.9$, df $= 1$, $p = 0.2$; Fig. 4.3). However, an analysis based on a different measure of long-term success – the mean number of fledglings produced per year of pair bonding – revealed that in years where all breeding attempts had failed, the more successful pairs (generally 'successful' but at present 'non-parental') stayed closer together than less successful pairs, particularly during resting (Fig. 4.3b). In contrast, when the same pairs were 'parental' in another year, we found the opposite pattern: although there was no significant difference in the overall within-pair distance (Fig. 4.3c), the 'parental' pair partners tended to rest farther apart than those which had raised fewer young to fledging (Fig. 4.3a). Thus, proximity between successful breeders may indicate a strong pair bond (Trillmich 1976; Anzenberger 1983) in the absence of goslings, whereas during brood care, spatial proximity between parents is likely to depend on the number of goslings. Therefore, an explanation for the finding that previously more successful parents stayed less close together compared with less successful pairs is that pair partners adjust their distance in such a way that they can oversee larger gosling groups, which are spread over a larger area, in order to guard them (Schneider & Lamprecht 1990; Spoon *et al.* 2006). Thus, for pair partners, a flexible, context-sensitive spatial distribution during parenting is probably an attribute of parental competence.

Throughout the rest of the year, a nearby mate is a social ally for access to resources, independent of reproductive status (Lamprecht 1986a, 1986b). This may add to an understanding of why unsuccessful pairs, such as those in our study, remained together for long periods of time, even in the absence of reproductive success or other factors such as hormonal compatibility (see Section 4.2). The decision to remain together may also be based on a similarity between male and female quality, as individuals more similar in quality tend to remain together rather than divorcing (McNamara *et al.* 1999). Furthermore, individuals may decide to remain paired due to a lack of available alternative partners or due to a failure in recognising the quality of alternative partners in the flock (Choudhury & Black 1993).

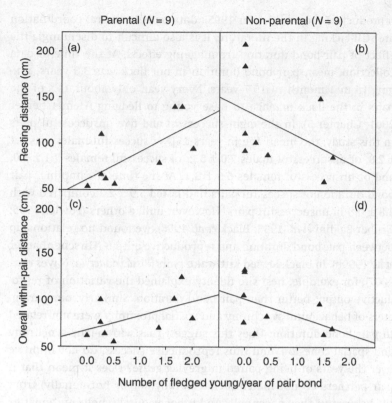

Figure 4.3   Relationship between fledging success (*x* axis) and mean resting and overall within-pair distances (*y* axis) in pairs of parental (*a*, *c*) and non-parental (*b*, *d*) greylag pairs. Each *triangle* represents a pair. Spearman's rank correlation: (*a*) $N = 9$, $r = 0.6$, $p = 0.08$; (*b*) $N = 9$, $r = -0.7$, $p = 0.04$; (*c*) $N = 9$, $r = 0.5$, $p > 0.1$; (*d*) $N = 9$, $r = -0.7$, $p = 0.05$.

The actual duration of a pair bond and its 'efficiency' in terms of reproductive success may depend on a number of factors, such as age (barnacle geese, *B. leucopsis*: Black & Owen 1995), physical partner compatibility (Choudhury *et al.* 1996) or social environment (Canada geese: Raveling *et al.* 2000). According to the 'mate familiarity hypothesis' (Black 2001), reproductive output should increase with the number of years of being paired. An increasing familiarity between mates may specifically affect the coordination of behaviours and reproduction. However, it is difficult to disentangle the effects of mutually adapting to one partner from a generally accumulated breeding experience. In many species with long-term pair bonds, experienced adults in newly formed pairs (due to divorce or loss of mate) have decreased reproductive success (Fowler 1995); data on behavioural coordination are still lacking in the literature. It is also difficult to disentangle the effect of pair bond duration from ageing effects. At the time of data collection, mean pair bond duration in our flock was 3.3 years, ranging from 1 month to 15.7 years. Every year, only about 11% of the pairs in the flock manage to raise young to fledging (Hemetsberger 2001; Chapter 5). In the eight successful and five unsuccessful pairs in this study, the mean age in years ±SD of successful males was 7.1 ± 2.8, of unsuccessful males 7.0 ± 5.2, of successful females 7.1 ± 3.0, and of unsuccessful females 6.6 ± 3.1. At the time of sampling, pair bond duration of successful pairs had lasted 5.9 ± 2.2 compared with 1.8 ± 0.8 in unsuccessful pairs. However, unlike others (Fowler 1995; Gruber-Baldini *et al.* 1995; Black *et al.* 1996), we found no relationship between pair bond duration and reproductive output (Hirschenhauser *et al.* 1999b). In black-legged kittiwake pairs (*Rissa tridactyla*: Naves *et al.* 2007), for example, nest site fidelity explained the variation of reproductive output better than pair bond duration. Similarly, our parameters of behavioural synchrony and spatial proximity were not related to pair bond duration. Does this suggest that within-pair synchrony and spatial cohesion, and thus reproductive success, do not improve over the years of being paired in greylag geese? Does it mean that if pair partners do not match (behaviourally and/or hormonally) from the beginning, they never will? And is behavioural synchrony a matter of compatible individual traits of the pair partners? These are some of the outstanding questions to be addressed in future studies of the proximate and ultimate substrates of pair bonding.

4.1.2 Leaders and followers

Another question we attempted to answer was whether the spatial cohesion within a pair results from the continuous one-sided efforts of one of the pair partners or from the degree of coordination between pair partners. Are some pair partners better 'co-operators' than others (Soares *et al.* 2010) and what would the costs and benefits of 'non-co-operators' be? To assess within-pair coordination, we studied in detail the 'leader and follower' roles of pair partners in five successful breeding pairs over two consecutive years: in the first year (2005) these pairs cared for young, in the second year (2006) they did not. In addition, we collected data from four pairs that had never successfully bred prior to the year of data collection. In the breeding season we recorded which of the mates initiated a change in location, change in direction, or a clear change in behaviour. We also recorded whether the other mate responded to this initiation (i.e. followed) or not. Unclear situations were not included in the analyses, such as when one or more goslings initiated a behaviour or change in location.

If within-pair cohesion resulted from the contribution of mainly one pair member, we would expect leaders and followers to emerge within a pair, not only within but also outside the reproductive season. A first analysis of male and female behaviour in the five successful pairs revealed that when with goslings, mothers led and fathers followed. If there were no goslings to care for, however, the roles were reversed: males were leaders and females followers. In other words, during brood care males followed their females more often than in the non-parental year (Wilcoxon signed ranks: $Z = -2.0$, $N = 5$, $p = 0.04$; Fig. 4.4), while in the non-parental year, females tended to follow their mates more than during the time when they were caring for young ($Z = -1.8$, $N = 5$, $p = 0.08$; Fig. 4.4). Thus, it seems as if both partners – rather than just one – adjusted to the situation and contributed to the maintenance of spatial cohesion. In the absence of young, the females were adjusting and contributing to pair bond maintenance by following their mates, probably in order to maintain the benefits of social support. These findings are in agreement with earlier studies on within-pair coordination in bar-headed geese (*A. indicus*: Lamprecht 1991, 1992), where departures were initiated either by males or by females, depending on the seasonally different demands of reproductive phases. In conclusion, leadership within the pair is largely dependent on functional needs.

Although, for biparental birds, coordination between pair partners is pivotal for successfully raising offspring (Fraser *et al.* 2002), in geese a certain degree of within-pair coordination should also be required for long-term pair bond maintenance in the absence of young. To assess the degree of within-pair coordination, we correlated the number of initiation events with the number of subsequent events within a pair for each observation. This resulted in a 'coordination factor' for each pair (Hirschenhauser *et al.* 1999b). To include only data collected in the absence of offspring we compared coordination factors from unsuccessful breeders (i.e. 'unsuccessful' over the entire pair bond duration) with those from the non-parental successful pairs ('successful' pairs in the past, but 'non-parental' in the year sampled). Even if limited by sample size, coordination factors for successful pairs without current parental duties (0.8 ± 0.04, $N = 5$ pairs) were slightly higher than in unsuccessful pairs (0.7 ± 0.1, $N = 4$ pairs). Again, the

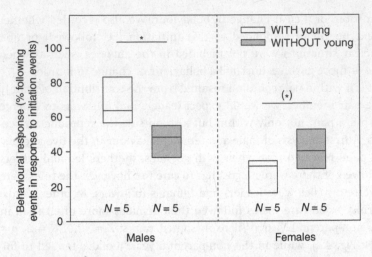

Figure 4.4   Males responded ('followed') more often to the females' behaviour during the year with brood care (with young: *open boxes*) than during the non-parental year (without young: *filled boxes*). Females more often followed a behavioural change initiated by the male in the non-parental year. *Boxes* represent second and third quartiles, *error bars* indicate 95% confidence intervals. *Asterisks* indicate significant differences: *$p < 0.05$, (*) $p < 0.1$.

behavioural coordination factors did not relate to the years of being paired ($r_s = 0.43$, $N = 9$, $p = 0.3$) or the female's age ($r_s = 0.5$, $N = 9$, $p = 0.2$). However, pairs with older males had a higher coordination factor ($r_s = 0.63$, $N = 9$, $p = 0.07$) than pairs with younger males.

To summarise, we observed that behavioural parameters at the pair level indeed correlated with pair quality as measured by reproductive success. Important features were a flexible adjustment between partners in functionally equivalent behaviours, spatial proximity and the 'within-pair coordination'. These behavioural markers of within-pair synchrony were influenced by the presence or absence of offspring, and possibly by the gander's age. It remains to be tested in the future whether older males may become better at adjusting to the female partner's pace of breeding and to disentangle the effects of age, experience and previous success in the male and the female partner. However, pair partners need to be matched from the start, and they indeed seem to be, as we found no indication of an improvement in the relevant behavioural parameters with time spent together. This highlights the importance of mate choice for becoming a successful pair.

4.2 ANDROGEN CO-VARIATION BETWEEN PAIR PARTNERS: AN INDICATOR OF MUTUAL SOCIO-SEXUAL ATTRACTION?

Seasonal regulation of breeding activities is essential for geese; for example, for the timing of hatching with respect to food availability, or for synchronising the end of the parents' moult with fledging of the offspring (Lorenz et al. 1979). The seasonal patterns of sex hormones, such as androgens in males and females, and oestrogens in females, peak around mating and drop over the subsequent breeding phases (Dittami 1981). The seasonal testosterone patterns in geese are typical for monogamous and biparental birds (Wingfield et al. 1990). Proper immunoassays based on group-specific antibodies (Möstl et al. 2005) and strict evaluation procedures of non-invasively measured hormone metabolites excreted in faecal droppings (Krawany 1996; Hirschenhauser et al. 2005) allowed us to study the seasonal patterns of various steroid hormones in individual geese in the Grünau flock in detail. In both sexes we observed a wide variation of excreted testosterone metabolites between individuals, in particular during the egg-laying and incubation phases. We first showed that seasonal patterns of testosterone varied due to social status in geese of both sexes. For example, male singletons had longer phases of high testosterone levels than males engaging in nesting activities with a partner (Hirschenhauser et al. 1999a). This is in agreement with the idea that prolonged phases of high testosterone are incompatible with paternal care (Hegner & Wingfield 1987; Hirschenhauser & Oliveira 2006). However, female geese also had a distinct seasonal testosterone pattern (Hirschenhauser et al. 1999a, 1999b; Fig. 4.5). Androgens in females are of adrenal origin and are to some extent produced in adipose tissue and growing ovarian follicles (Johnson 2002). Female androgens are related to egg production and fecundity, with the highest peaks being found in females from socially monogamous species (Ketterson et al. 2005).

We observed that in some pairs individual testosterone patterns of male and female pair partners took a more 'parallel' course than in others (Hirschenhauser et al. 1999b). The degree of within-pair testosterone compatibility (or testosterone co-variation) (TC) was quantified as a correlation coefficient between the male and the female partner's seasonal testosterone, which covered the entire range between 'minus-one pairs' (hormonally non-matched pair partners) and 'plus-one pairs' (hormonally matched pair partners). As described earlier, every year only a small fraction of all pairs in the studied

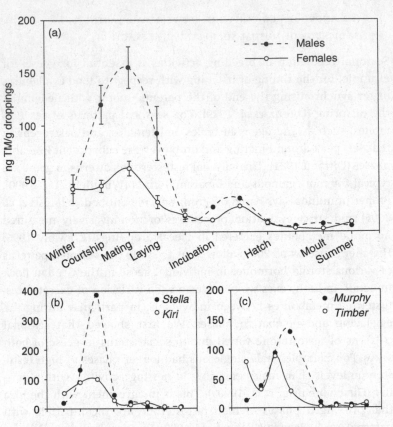

Figure 4.5 Seasonal patterns of excreted immuno-reactive testosterone metabolites (TM) in droppings of paired male (*filled symbols, dashed line*) and female (*open symbols, solid line*) greylag geese. (a) Mean TM (±SE) per seasonal phase. Seasonal phases were assigned by individual dates when females had laid the first egg and for the male partners respectively. Sample sizes (m/f) varied between seasonal phases: Winter – 18/20, Courtship – 19/20, Mating – 19/20, Laying I – 18/18, Laying II – 18/18, Incubation I – 1/1, Incubation II – 1/1, Hatch – 19/20, Goslings – 18/19, Moult–13/14, Summer – 18/19. (b) Example of the individual TM patterns of a testosterone co-variant (TC)-matched pair (TC = 0.6, p = 0.02) and (c) of a non-matched pair (TC = 0.3, n.s.).

goose flock successfully raise offspring to fledging (Hemetsberger 2001; Chapter 1). We wondered whether TC was correlated with reproductive success. Although the causality remained open at that time, we observed that TC was correlated with reproductive output not only in the year sampled, but also for a whole lifetime. Females from pairs with a higher TC (i.e. from better-matched pairs) laid bigger clutches with on average heavier eggs. In addition, both partners had raised more offspring to fledging over the years than pairs with a lower TC (Hirschenhauser *et al.* 1999b). We assumed that the phenomenon relied on the male's responsiveness and ability to adjust to the female's decisions on when to start laying and incubating (Moore 1982; Caro *et al.* 2009). However, a process of increasing mutual adjustment over the years of being paired (i.e. becoming more and more synchronised) seemed unlikely, as TC was not correlated with pair bond duration. This mirrors the results concerning behavioural parameters of within-pair synchrony and coordination (see above). For example, both pairs with the longest pair bonds (18 and 11 years, respectively) had non-significant TC coefficients (Hirschenhauser *et al.* 1999b). It was unclear whether and how the female or both partners were choosing a compatible, responsive partner during pair formation, or whether the patterns we saw were the result of random matching.

Ten years later we sampled 21 new greylag pairs to check our previous results by this replication. Again, TC co-varied with fitness parameters, notably fledged offspring (Weiß *et al.* 2005). The detailed lifelong records of each flock member kept since 1990 (see Chapter 1) enabled us to perform a comprehensive analysis of various life history parameters, which included the seasonal testosterone profiles of 44 male and female pair partners (i.e. the 23 pairs from the initial study and the 21 new pairs), as well as 7 trios (Weiß *et al.* 2010b). This larger-scale analysis indicated that the female's experience, for instance regarding the timing of breeding events, plays a central role in maintaining TC, as older females had higher TC with their partners, irrespective of the age difference between them. In contrast, TC was not a function of the male partner's age. We also confirmed that TC was not increasing over the years of being paired. On the contrary, it appeared that hormonal compatibility decreased with the years of being paired to the *same* partner ($r = -0.4$, $N = 43$, $p = 0.013$). Thus, partners that do not match from the start probably never will; and being TC-matched with a partner at a certain time period is not a reliable 'prognosis' that it will always remain this way. There may be some form of 'attritional effects' over time even in greylag partnerships. This triggered new questions; for example, whether TC was indeed a matter of mate choice, and which behavioural parameters were used for assessing each other's responsiveness. Is individual style the cue for choosing a particularly compatible partner, and which suites of individual attributes is it based upon? In fact, the attempt to unravel the causal link between successful reproduction and hormonal partner compatibility amounts to a 'chicken and egg' dilemma.

Some useful insights were gained from experimental work with domestic geese (A. domesticus: Hirschenhauser et al. 2010). They preferably form small harems year-round; normally a dominant male is paired with a variable number of females. Domestic ganders also engage in parental care as do their wild ancestors, and they vigorously guard their female(s) and nest(s). For testing TC experimentally, we kept 16 male–female pairs of domestic geese for monitoring seasonal testosterone profiles throughout an entire year, thus shrinking harem sizes to one female – in essence, forcing individuals into monogamy. In many pairs, strong bonds were observed between partners (Fig. 4.6). Domestic geese produce large quantities of eggs over much longer laying periods than wild geese (on average 34 eggs in up to 3 months in the studied population), and domestic goose females do not necessarily engage in incubation. Hence, we were not surprised to find that clutch size, egg weight and breeding success were not correlated with within-pair TC in domestic geese. However, the laying period was shorter in females from pairs with higher TC, and being with a matched partner apparently increased the probability of actually incubating.

Comparing the seasonal testosterone patterns for one year, in which pair partners were assigned by us at random ('random pairs'), with another year, in which pairs were allowed to choose their 'preferred partners', shed new light on the male and female partner's contribution to TC: the female's seasonal testosterone patterns were modulated by mate choice. Particularly during the early laying phase, females with 'preferred partners' had higher seasonal testosterone levels than females from 'random pairs' (Fig. 4.7), confirming the general notion that high androgen levels in females are related to sexual motivation and fecundity (Ketterson et al. 2005), not only in birds (Carter 1992). This indicates that there are two episodic components of the TC phenomenon: the female's sexual responsiveness to the partner and the male's responsiveness to her readiness to breed (Hirschenhauser et al. 2010). Thus, both partners contributed their share in a matched seasonal timing. The within-pair TC, however, was not explained by partner preference, although we observed some pairs with remarkably well-matched TC in the domestic goose pairs (Fig. 4.8).

Social interactions or environmental challenges may be very stressful and effectively modulate individual behaviour and physiology (DeVries et al. 2003). Activation of the hypothalamic–pituitary–adrenal

Figure 4.6   Sequence of a domestic goose pair engaging in biparental (nest) care: after the female (*left*) had laid an egg, the male (*right*) participated in covering the clutch with nesting material.
© K. Hirschenhauser

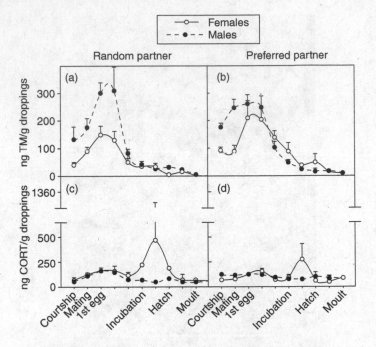

Figure 4.7   Seasonal patterns of excreted immuno-reactive testosterone
(TM: *a*, *b*, *top row*) and corticosterone metabolites (CORT: *c*, *d*, *bottom row*)
from droppings of domestic geese when they were kept in different
social settings. Throughout a full breeding season geese were kept
as male–female pairs either with (*a*) a randomly assigned partner
(*Random partner*) or (*b*) a self-chosen partner (*Preferred partner*). During
the early laying phase, females with a preferred partner had higher TM
than females from random pairs (*a*, *b*). Incubation is a metabolically
demanding time for females. The peak of stress hormone metabolites
(CORT) typical for females towards the end of incubation was
dramatically higher in females with a random than with a preferred
partner (*c*, *d*). Seasonal phases as in Fig. 4.5; males are *dashed lines* and
filled symbols, females are *solid lines* and *open symbols*. Part of the figure
(for TM) reprinted from Hirschenhauser *et al.* 2010 with permission
from Elsevier.

(HPA) stress axis induces a cascade of physiological changes including the release of glucocorticoid hormones from the adrenal glands.
These stress hormones induce certain behavioural and physiological
facets of the (anti-)stress response (Sapolsky *et al.* 2000). In the greylag geese, we monitored the seasonal baseline patterns of excreted
immuno-reactive corticosterone metabolites (CORT) from 32 males

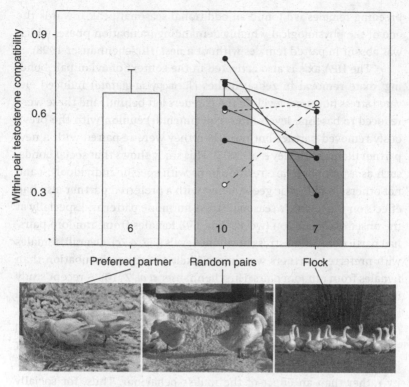

Figure 4.8   Within-pair testosterone compatibility (TC) of random
pairs (2005) and when kept in one flock in the following year (2006).
Numbers inside bars are *N* pairs. Dependent data (from the same pairs in
different years) are connected with lines. In the flock five of the seven
pairs had reduced TC (*filled dots, solid lines*), one pair's TC was stable,
one pair increased after reunion of the flock (*open dots, dashed lines*).
Part of the figure (dependent data) reprinted from Hirschenhauser
et al. 2010 with permission from Elsevier. Photographs © K.
Hirschenhauser.

and 24 females (Hirschenhauser 1998). The seasonal CORT peak of
ganders differed due to social status, and thus due to the seasonal
requirements of reproduction; during the mating season singletons
had higher CORT than paired males, whereas after hatching of goslings the fathers ('parental males') had higher CORT than all other
males (Kotrschal *et al.* 1998a). Paired males and females had a similar seasonal timing, with a CORT peak during the pre-laying phase
and a trough at the end of summer. However, we observed no meaningful within-pair patterns based on seasonal baseline CORT. Among

breeding females we found an additional seasonal peak towards the end of the physiologically highly demanding incubation phase, which was absent in paired females without a nest (Hirschenhauser 1998).

The HPA axis is also activated in the context of avian pair bonding; mate removal in zebra finches (*Taeniopygia guttata*) induced elevated stress hormone levels in the partners left behind, and these were reduced to baseline levels after experimental reunion with the previously removed partner but not when they were re-paired with a new partner (Remage-Healey *et al.* 2003). This study shows that social bonds, such as a pair bond, are relationships with specific individuals – and not others. In domestic geese, being with a preferred partner also had effects on the female's seasonal stress hormone patterns. Especially at the onset of incubation (weeks 3 and 4), females from 'random pairs' had particularly high stress hormone levels (Fig. 4.7c). Overall, females with preferred partners were more likely to initiate incubation than females from random pairs (Hirschenhauser *et al.* 2010). A recent study in socially monogamous Gouldian finches (*Erythrura gouldiae*) reported that females rapidly responded with systemic corticosterone increases to being paired experimentally with a male of low quality. The rapid nature of the stress hormone response in these finches suggests that it was due to the female's initial perception of the partner's quality rather than an effect of the male's behaviour. Thus, for socially monogamous birds with long-term and continuous pair bonds, partners may indeed rapidly assess a mate's quality during pair formation. Consequently, females paired with low-quality males maintained high stress hormone levels for several weeks and started egg-laying after a delay (Griffith *et al.* 2011). These examples, as well as our goose studies, suggest that being with a matched partner is crucial for breeding efficiency.

4.3 SOCIAL MODULATION OF HORMONAL CO-VARIATION

Throughout a second year, the domestic geese from the 'random pairs' group (see above) were kept as one large flock, and their seasonal testosterone profiles were monitored (Hirschenhauser *et al.* 2010). Although the median degree of TC did not change significantly, in the majority of pairs, TC was reduced to some extent when the geese were kept in a flock (Fig. 4.8). This is best demonstrated using individual case studies. For example, we had two previously well TC-matched pairs that merged into a four-member group when kept as a flock in the second year (Fig. 4.9a). It seems that when in the flock, especially

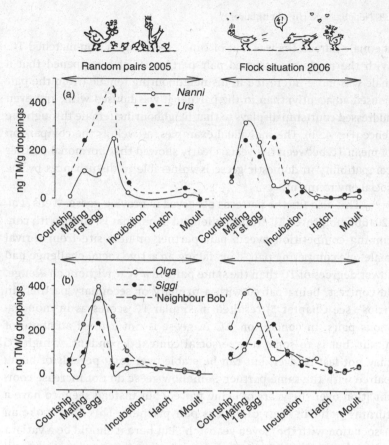

Figure 4.9   Individual examples of seasonal patterns of excreted immuno-reactive testosterone metabolites (TM) in domestic goose pair partners. (a) The pair partners *Nanni* (female) and *Urs* (male) were well TC-matched when kept as a pair in the 'random pairs' year (TC = 0.7, $p < 0.001$). However, when they were freely moving and interacting in one large flock in the following year, the timing of seasonal TM peaks of both partners was very different from the previous year, and in that year TC was also reduced (TC = 0.4, n.s.). (b) *Olga*, *Siggi* and 'Neighbour' *Bob*: the female *Olga* and her (randomly assigned) male partner *Siggi* were relatively well TC-matched (TC = 0.6, $p = 0.002$) when they were kept as a pair. However, the male of the pair that was housed adjacently and separated by a wire mesh fence ('Neighbour' *Bob*) frequently engaged in courtship (e.g. angled neck) displays directed towards *Olga* – through the fence. We also plotted this male's seasonal TM for the random pairs year, as well as for the following year in the flock. *Olga* and her 'fancy man' *Bob* were well TC-matched in both years (TC = 0.9, $p = 0.004$ and TC = 0.5, n.s., respectively), while *Olga* and *Siggi*'s hormonal synchrony was diminished when they were freely interacting in the flock (TC = 0.5, n.s.). *Filled symbols* and *hatched lines* are male testosterone metabolite patterns, *open symbols* and *solid lines* are female androgen patterns.

the male named *Urs* was 'out of tune', resulting in non-matched TC. With the randomly assigned pair partners it also happened that a male was more interested in his neighbouring female (from the pair housed adjacently) than in the partner it was housed with, and even addressed courtship displays to that neighbouring female through the fence (Fig. 4.9b). The individual examples, as well as the comparison of mean TC between the years, clearly showed that hormonal partner compatibility in domestic geese is vulnerable to disturbances by the social environment.

This also fits the patterns observed in greylag geese. Weiß *et al.* (2010b) monitored TC from challenged pairs; that is, those with continuing competition over a pair partner or nest site from a rival male or a competing pair. Pairs facing an active social challenge had lower degrees of TC than the same pairs in a year without challenge. In contrast, being paired with a primary or secondary partner (in 'trios', see Chapter 5) resulted in similar TC scores as in monogamous pairs. In conclusion, TC in geese is not a fixed attribute of a pair, but is to some degree social context-dependent. A high TC may not last forever, but can be stable over long periods of being paired with the same partner. Somehow geese do not increase coordination over the years of being paired, but instead seem to have a turning point in their careers as pair partners. This may evoke an association with the 'seven-year-itch', but here it should be a variant called the 'nine-year-itch': the longest-lasting pair bond that was still TC-matched at the time of sampling was 8 years (observed in three pairs), while TC was no longer significant in any of the pair bonds that had lasted 9 years or more (Weiß *et al.* 2010b). As the coordination and compatibility of previously matched pairs was substantially affected by the stability of the social environment, we should be aware of the fact that TC reflects a pair's status quo, with TC being affected by internal (individual), dyadic and external (embedding of the pair in the social web) factors. As such, TC may be an indicator of mutual socio-sexual attraction, which is an important basis for optimal cooperation within a pair, but of course it is not the only factor involved. Taking together the behavioural and hormonal within-pair patterns, we may conclude that in the long term it takes more than two compatible partners to make a 'good pair'. Selection has probably favoured pairs in which both partners cooperate in a continuing process of mutual responsiveness, seasonally changing assignments and multilateral interactions with their social environment.

SUMMARY

The formation and maintenance of a pair bond are the principal mechanisms for optimising reproductive success in monogamous systems. It is still puzzling, however, why some pairs remain paired with each other even if they have no reproductive success. One approach to studying pair bond mechanisms is to focus on either the female or the male partner. In our approach, however, we shift the unit of analysis from the individual to the pair, which we term 'partner compatibility'. This chapter summarises long-term data on the compatibility of both male and female pair partners' seasonal fluctuations in behaviour and testosterone levels. Detailed time budgets of male and female pair partners unravel seasonal and context-dependent patterns of alternating behavioural roles and of the optimal within-pair distance. We show that 'flexible' behavioural adjustment to the partner plays a role and that leadership among pair partners is basically dependent on functional needs. Thus, in goose pairs there is active mutual responsiveness between pair partners, which may vary seasonally. Similar patterns have emerged from detailed studies of the pair partners' seasonal testosterone compatibility (TC). Pairs with a higher degree of TC have larger clutches with bigger eggs and greater lifetime reproductive output than pairs with a lower degree of TC. This finding was replicated in a second data set. An overall analysis of all hormonal pair data shows that TC does not improve with the years of being paired and is also not a function of age of either pair partner. In experiments with domestic geese, we find the entire range of TC-matched and non-matched pairs even in these polygynous and biparental birds. However, the hypothesis of a mate-choice effect on TC was not confirmed with experimental tests. Tests with both greylag and domestic geese reveal that a pair's TC and compatibility is vulnerable to disturbances from the social environment. Partner compatibility seems to be a crucial feature of pair bond quality and the pair's reproductive success; however, in the long run, it takes more than merely two compatible partners to be a successful pair. Selection seems to have favoured pair partners with continuing mutual adjustment of behaviour and reproductive physiology.

5

Alternative social and reproductive strategies

In the majority of birds, the prevailing social strategy is the formation of monogamous, heterosexual pair bonds (Emlen & Oring 1977; Björklund & Westman 1986). In some species, including geese and swans, these pair bonds are formed for many years and often for life (see Chapters 1, 2 and 4). Accordingly, heterosexual pairs, with or without offspring, make up the majority of members in a goose flock, but these are by no means the only type of pair bond observed. In the early years of goose research, scientists had already noticed the formation of male–male pairs and of 'pairs' that obviously included three or even more individuals (Lamprecht 1987; Huber & Martys 1993; Chapter 2). Such bonds are usually rare, but are nonetheless regularly observed. As with heterosexual pair bonds, they are characterised by close proximity between partners, greeting behaviour and the triumph ceremony (see Chapter 11), a lack of agonistic interactions within the unit, and behavioural synchronisation (Chapter 4), and may be as stable and long-lasting as heterosexual bonds. Most reports of alternative pair bonds stem from captive or semi-natural conditions, but they have also been observed in the wild, indicating that – at least from a qualitative viewpoint – such relationships are also representative for natural populations. The identification and quantification of alternative pair bonds, however, is more difficult in wild, unmarked populations, which may be one reason for the predominance of reports from captive and semi-natural populations. Importantly, the majority of individuals engaged in alternative pair bonds form monogamous, heterosexual bonds at earlier or later

The Social Life of Greylag Geese: Patterns, Mechanisms and Evolutionary Function in an Avian Model System, ed. I. B. R. Scheiber *et al.* Published by Cambridge University Press. © Cambridge University Press 2013.

stages in their lives (Kotrschal *et al.* 2006), illustrating that these bonds are not exclusively a specialist strategy.

5.1 INDIVIDUAL LIFE HISTORIES

Many individual life histories in our flock illustrate the flexibility in bonding types. Some are described here, with a note of caution for the reader not to assume an individual's sex from its nàme.

Darwin, a male raised by his parents in 1992, formed a pair bond with a 2-year-old female, *Lauser*, in the spring of 1995. In summer 1999, another male, *Laura*, who had recently lost his male pair partner, joined the pair. They formed a trio for several years in which *Darwin* and *Lauser* formed the obvious core unit. *Laura*, who showed little or no interest in *Lauser*, was *Darwin*'s secondary partner, with the exception of the breeding season, when he was vigorously chased away by *Darwin*. During a January storm in 2004, *Darwin* and *Lauser* disappeared, together with several other geese. *Darwin* returned 3 months later with another female, *Noodles*. *Laura* immediately rejoined the new pair. As before, *Laura* was only interested in, and bonded with, *Darwin*, but *Noodles* was *Darwin*'s main partner and showed rather frequent aggression towards *Laura*. *Noodles* disappeared in the winter of 2007 and *Laura* became *Darwin*'s main and only partner. Three months later, *Darwin* broke up with *Laura* and formed a pair bond with a young male. *Laura*, now 18 years old and with little chance of having another 'serious' pair bond, remained alone for half a year until he loosely bonded with another male of the same age as himself. The two old males maintained their loose bond until *Laura*'s death early in 2010. *Darwin*'s new partner disappeared during the moult in 2009, and *Darwin* remained alone until his death in spring 2011.

The male *Keiko* was hand-raised in 1996 and remained tightly bonded to his siblings until he was almost 2 years old, when the sibling bond to his brother *Skana* turned into a tight pair bond. The pair of brothers resembled the 'classical' male–male pair, as described by Huber and Martys (1993), in every respect: both males were particularly loud and aggressive. *Keiko* and *Skana* were frequently accompanied by young females with whom they copulated, but the brothers formed no stable social bonds with any of their female sexual partners (see also Kotrschal *et al.* 2006). In spring 2002, which was 4 years after their pair formation, *Skana* disappeared and *Keiko* loosely reattached to his sisters and foster mother for several weeks. After a year, *Keiko* formed a loose pair bond with the female *Jana*, but also occasionally courted other males. *Keiko* and *Jana*'s bond eventually became more stable and the pair had several successful breeding attempts, but did not manage to raise their young to fledging. In autumn 2004, the female *Jaspis* started following *Keiko* and became *Keiko*'s secondary partner. *Jana* tolerated *Jaspis*' presence and continued to be *Keiko*'s prime focus of attention until her death in the autumn of 2006. *Jaspis* subsequently assumed *Jana*'s role as *Keiko*'s main partner until *Keiko*'s death in 2010.

5.2 MALE-MALE PAIRS

Pair formation between same-sex partners is a common phenomenon in animals under captive or semi-captive conditions, but also in the wild (Sommer & Vasey 2006). In geese, homosexual pair bonds occur exclusively between males, and never between females (Kotrschal *et al.* 2006). Female–female pair bonds do occur in other species of birds (e.g. pukekos, *Porphyrio porphyrio*: Jamieson & Craig 1987; Eurasian oys-tercatchers, *Haematopus ostralegus*: Heg & van Treuren 1998) and we can only speculate about why they seem to be absent in geese. However, when having a choice of pair partners, males will be the more power-ful, higher-ranking allies than females, and when suitable male part-ners are not available, the long-term kin bonds among female geese (see Chapter 6) may prove to be a better alternative than pairing with another female.

Male–male bonds are frequently alternated with other types of pair bonds (see individual life histories above). In our flock, 49% of the males engaged in heterosexual pair bonds with females only, while 37% of the males formed pairs with both males and females over the course of their lives. The remainder of the males (14%) exclusively paired with other males (Kotrschal *et al.* 2006). Male-paired males show typical male behaviour (Huber & Martys 1993) with the occasional exception. When two males attempt copulation, both males usually try to get the top position (Fig. 5.1), which often ends the attempt and can even lead to serious fights between the partners. Sometimes, however, one male will adopt the female position, or they avoid the dilemma altogether by copulating with an unpaired female instead. In the lat-ter case, the post-copulatory display is then directed towards the male partner. Male-paired males, however, copulate less frequently than heterosexual pairs (Kotrschal *et al.* 2006). The fact that male-paired males behave like males in almost every aspect sometimes also leads to these pairs being overly aggressive and loud, as shown in the exam-ple of *Keiko* and *Skana* above. Such behavioural conspicuousness was

Figure 5.1   Two male greylag geese attempting to mount one another. The situation develops into mutual biting and the copulation attempt is eventually aborted. © B. M. Weiß.

long assumed to be typical of male–male pairs (Huber & Martys 1993) but, on taking a closer look, this seems to be the exception rather than the rule. This impression is corroborated by recent findings, which showed that homosexually paired males are generally less aggressive and lower-ranking than heterosexually paired ones (Weiß *et al.* 2011; Chapter 7). This may well be due to a certain imbalance in the strength of affection in many male–male pairs, where it seems that one partner is actively pursuing the pair bond, while the other is merely tolerating it. Such pairs frequently involve males who have relatively recently lost their female partner and show only low agonistic motivation. The fact that widowed, older males are likely candidates for male–male pairs is reflected in the increase in male–male pairs with pair bond number: about 20% of all first-time pair bonds are male–male, while this percentage steadily rises to 80% for the sixth pair bond (Kotrschal *et al.* 2006). A lack of female partners certainly plays a role in this increase in male–male pairs with age, as females are more frequently preyed upon and are hence shorter-lived than the males. Along the same lines, heterosexually paired males were younger (2.3 years) at first pair bond than male-paired males (3.7 years), which may indicate that these males had some difficulty in obtaining a female partner (Kotrschal *et al.* 2006). Indeed, the frequency of male–male pairs increases when the sex ratio of the flock becomes increasingly male-biased (Huber & Martys 1993). However, it must be noted that they also occur when the sex ratio is balanced (Kotrschal *et al.* 2006).

These observations illustrate that the motivation for male–male pair formation can be quite diverse. There are a small number of males who engage in long-term stable bonds with males from an early age onwards and frequently form symmetrical, loud and aggressive pairs. Such pairs often last for life, and the male surviving his partner will very likely re-pair with another male. These males are probably best described as 'homosexual', while 'homosocial' may be a more appropriate term for the majority of male-paired males (Kotrschal *et al.* 2006). These homosocial males are typically older males that have lost one or more female partners and have little chance of finding a suitable female partner soon. Such males will frequently form a loose association with another male sharing this fate, or give in to the advances of homosexual males. Furthermore, the homosocial males include young males who cannot yet successfully compete for females, such as many youngsters of the 2004–6 cohorts. In those years, female offspring were scarce and those available were paired as soon as they left the family unit, or even before that time. Quite a number of the young males subsequently formed loose male–male bonds that almost all broke up again in 2008 and 2009, when more young females became available.

As male–male pairs do not produce and raise offspring, such pairs have often been considered as dysfunctional and maladaptive, but certain benefits of male–male pairs immediately become apparent, particularly for the homosocial ones. Most notably, a pair partner is a highly useful social ally. Although male-paired males are frequently lower-ranking than heterosexually paired ones, paired individuals are clearly dominant over singletons and thereby have better access to resources (Weiß *et al.* 2011; Chapter 7). This, in turn, may improve the males' chances of survival and/or of successfully competing for a female in the future. Male–male associations could thus be seen as some form of coalition. Male coalitions to gain a reproductive advantage are common in vertebrates (Whitehead & Connor 2005); however, we have no indication that greylag males associate with other males to actively gain access to females. Accordingly, homosocially paired males may make the best of a bad job (Dawkins 1980) and engage in a same-sex pair bond to remain high-ranking while waiting for better reproductive opportunities to arise.

These arguments are not applicable to the minority of males who never engage in heterosexual pair bonds. Still, male-paired males may have direct reproductive success via copulation with unpaired females or by producing extrapair young. With the introduction of molecular methods, extrapair young in monogamous bird species have been shown to be more frequent than previously thought, and thus may be a hitherto unrecognised source of reproductive success for homosexual males. This may be particularly true for the aggressive and showy type of males, as high-ranking males have been shown to be preferred extrapair partners in other birds (black-capped chickadees, *Poecile atricapillus*: Otter *et al.* 1998). Indeed, as indicated earlier, male–male pairs seem to be fairly attractive, at least to young, unpaired females, who we frequently observe to be sexual partners of homosexually paired males. Based on genetically determined parentage, we were recently able to identify the fathers of six offspring hatched from unpaired females; two of these were indeed fathered by homosexually paired males. Interestingly, the two fathers had formed a strong homosexual (rather than homosocial) pair bond with each other and produced one offspring each in the same year with the same female. However, neither of the two goslings survived to fledging, which is a common occurrence when females raise young without a male partner (Schneider & Lamprecht 1990). Fathering offspring of unpaired females will thus only convey fitness benefits in exceptional cases. Furthermore, we found no evidence for extrapair fertilisations involving paired females in our preliminary data set and, therefore, there is little support for direct fitness benefits of male–male pairs. However, less than 20% of heterosexual pairs ever raise young to fledging and, despite low direct fitness effects, the formation of homosexual pairs may be the better option for some males. Unfortunately we cannot tell how successful a male would have been had he formed a heterosexual bond instead.

5.3 POLYGAMOUS PAIR BONDS

The vast majority of polygamous pair bonds are in the form of trios, which typically form through the secondary attachment of a male or female to one member of an already existing pair, male–female or male–male. In most cases the core pair remains distinct because the now polygynously paired individual continues to allocate most of its pair bonding behaviours to its initial, primary partner and less to the newly attached, secondary partner. The primary and secondary partner in a trio typically tolerate one another but seldom direct any other pair bonding behaviours towards each other, particularly when two females are paired to the same male. Between 1994 and 2009, about

10% of the pair bonds formed were polygamous, with roughly half of them in the form of a male paired to two females, and the other half consisting of two males and one female. We also observed one case of three young males pair bonded with each other, and one harem consisting of a male paired with four females simultaneously. As in trios, the primary female was clearly discernible, and so was the secondary female, who received less attention than the primary female but more than the other two. This rate of polygamy closely resembles that observed in a semi-captive flock of bar-headed geese (*Anser indicus*), but only polygynous males with one to four additional female partners were reported; this may be an artefact of an unusual female bias in this particular goose flock (Lamprecht 1987). In contrast, the sex ratio of our flock has varied from equilibrium to a male-biased sex ratio. Unlike in bar-headed geese, however, we have never detected a relationship between the number of trios formed and the sex ratio of the flock in any given year (female–female–male trios: $r_s = 0.003$, $N = 16$, $p = 0.99$; female–male–male trios: $r_s = 0.148$, $N = 16$, $p = 0.583$). The attachment of a third party to an existing pair was considerably less frequently observed in wild barnacle geese (*Branta leucopsis*), where polygynous pair bonds accounted for only 0.4% of the total number of pair bonds observed (Black *et al.* 2007).

Even more than male–male pairs, trios seem to fall into different categories depending on the sex of the trio members and the formation history. Female–female–male trios typically consist of a male bonded to a primary female; a secondary female later joins the pair. The two females tolerate one another to some extent but direct all bonding behaviour towards the male and not to each other. This situation is somewhat different in about one-third of the female–female–male trios, in which the two females are sisters that have retained their family bonds (Fig. 5.2; Weiß *et al.* 2008; Chapter 6). In those cases, a male courting one of the sisters frequently finds himself paired to both of them in the end. Also, we frequently observe cases in which one sister, after having lost her own partner, joins a sister and her partner. In either case, the male copulates with both partners, and quite often the secondary female seems to be more active in soliciting copulations than the primary female. If both females nest in a given year, however, the male almost always follows and guards his primary female, particularly if that female successfully hatches young. During this time the secondary female is virtually left on her own and can only obtain appreciable support from the male if she manages to closely coordinate her own breeding to that of the primary female,

Figure 5.2   A female–female–male trio performing greeting behaviour towards each other. The trio is composed of a male (*centre*) paired to the female on the *left*, whose sister (on the *right*) joined the pair after losing her partner to a fox. © B. M. Weiß.

both in time and space. We observed one case in which two sisters succeeded in this close coordination and raised young together with one male. This resulted in a 'superfamily' of 12 fledged offspring between the three parents (Weiß *et al.* 2008), which fulfils the definition of a *crèche* (Eadie *et al.* 1988) and shows that polygamous pair bonds may be highly successful under certain circumstances.

Nine of the 14 female–female–male trios observed between 1994 and 2009 were terminated by the death of one of the female partners, in which case the remaining male and female continued as a monogamous pair. Three times the secondary female formed a monogamous pair bond with another male, and once the male divorced both female partners. We only observed once that the primary and secondary females swapped roles. In the latter case the original primary female was presumably infertile, while the secondary female repeatedly raised young to fledging. Nevertheless, the initial primary female remained closely bonded to her partner until her death.

Matters become even more complex when it comes to female–male–male bonds. These can come about when a male tries to court a paired female, and neither the female's original partner nor the competitor manages to drive the other one off. After high initial levels of aggression, such situations often end in a truce that can even extend to a certain amount of socio-positive behaviour between the two males. The secondary male is typically chased away vigorously by the primary male during the mating season, but may reattach to the pair again in summer. In the five trios formed that way over the last 15 years, one primary male successfully disposed of the secondary male after several years, while another time the secondary male superseded the female's original partner. Twice, the secondary male eventually formed a pair bond with another female, and once the initial competitors formed a homosexual pair after the female's death.

The formation of a trio follows a similar path when a male attaches himself to a heterosexual pair, but is clearly interested in the male of the pair. Under these circumstances, the female and the secondary male interact on a continuum from hostility to friendliness, but in essence the female is paired to her initial partner only. All five cases of such trios between 1994 and 2009 ended with the death or disappearance of one or more of the trio members, after which the remaining partners continued their bond monogamously. Finally, we are aware of two cases in which a male attached to a heterosexual pair, but it was unclear whether he directed his attention towards the female or the male. In both cases, the secondary male dropped out of the bond again after about a year.

In addition to a male attaching to a heterosexual pair, female–male–male trios may also form when a female attaches herself to a homosexual pair. This seems to be a strategy typical of young females, but only a few of these have been well accepted into the social bond, rather than simply being copulated with, if the males adopt the strictly homosexual lifestyle. We only observed one such successful attachment in 15 years, but that female eventually broke off again and later formed a monogamous pair bond with a heterosexual male. On the other hand, homosocial males may readily accept females into their social unit and may even actively court them. Two such cases are known from 1994 to 2009. Here, the trios persisted for years, but the male–male core shifted to a male–female core and the surplus male was eventually expelled. Notably, in both cases, the latter had been the one to initiate the homosexual pair bond, while the tolerating individual ended up with the female.

The different types of trios illustrate that the motivation for trio formation can be diverse, and accordingly the potential benefits have to be viewed from a number of different angles. For this purpose it should be kept in mind that trios are primarily an alternative bond for the attachee, rather than for either member of the core pair. This does not preclude the pair from gaining potential benefits from such an alliance, but the polygamously paired individual cannot actively achieve this. Lamprecht (1987) studied the costs and benefits of polygynous pair bonds in a semi-captive flock of bar-headed geese and found that secondary females spent less time feeding and more time alert than primary females. Also the reproductive success of secondary females was lower than that of primary ones, but higher than that of unpaired females. Primary females were dominant over secondary ones and the reproductive success of the primary female was not affected by the presence of a secondary female, indicating that secondary females are not significant competitors of primary females. Lamprecht's data also hinted at the possibility that higher-ranking and more experienced males represent preferred partners for secondary females. He further speculated that secondary females may not only enhance their reproductive success relative to singleton females, but they may also be more successful as a secondary female paired to a high-ranking male than as being the only female of a low-ranking male. This line of reasoning closely resembles the rationale of the polygyny threshold model (Orians 1969), which assumes that a female can be more successful if she mates with an already paired male in a better habitat than with an unpaired male in a poor habitat, despite the prospect of losing some or all of the male's assistance in rearing the young (Björklund & Westman 1986). Polygamous bonds in bar-headed geese thus seemed to be an attractive alternative for secondary females at hardly any costs to the primary female and, in fact, increased reproductive success for the polygynously paired male.

Indeed, this picture is not too different in our greylag goose flock: over the course of 15 years, the monogamously paired females raised a mean of 1.3 offspring to fledging, primary females a mean of 1.2, and secondary females a mean of 1.1 offspring. In contrast, only one unpaired female was successful during that time, resulting in a mean of only 0.1 fledged offspring per unpaired female in the same time frame. Hence, being a secondary female is certainly more beneficial in terms of reproductive success than being unpaired. On the other hand, reproductive success did not seem to differ significantly between monogamously paired females, primary females and secondary females. Only 10 of the 14 female–female–male trios persisted through a breeding season though; hence potential differences – or the lack thereof – might become more apparent as sample sizes increase. Furthermore, it should be noted that the three instances in which secondary females fledged young included the superfamily, where the primary and secondary female were sisters (see above). In the other two cases, the primary female had not even attempted to nest and the male thus contributed somewhat to raising of the secondary female's young. Hence, secondary females were only successful if they had at least some male support, either through a reproductive cooperation with the (related) primary female or because the primary female failed. Unlike our greylag geese, Lamprecht's bar-headed goose flock lived in a fenced-in area, where females may have been less exposed to predation and therefore not as dependent upon the active participation of a male in protecting the young. Again, we will need larger sample sizes to investigate the role of male support for secondary females in more detail. In addition to direct fitness benefits, secondary females enjoy the benefits of a higher dominance rank and social support in a similar manner as homosocial males. This may allow these females to be in a better condition for reproduction once they procure their own suitable partner than previously unpaired females. In addition, female mortality is high during the breeding season and secondary females always inherited the primary female's position in the case of the latter's death (see also Lamprecht 1987).

The benefits of males joining a pair in pursuit of the female are similar to those for secondary females. Being part of a trio allows males to maintain a higher rank than singletons. This may facilitate their competition for a partner once a suitable candidate appears. A secondary male may also successfully displace a female's former partner or might gain reproductive success by sharing paternity with the primary male. This, however, would represent a clear cost to the primary male and most probably accounts for the temporary break-up of many female–male–male trios during the breeding season. None of the female–male–male trios that did persist throughout the breeding season successfully raised young to fledging, but our DNA-based determination of parentage revealed that secondary males do fertilise eggs. In fact, all three goslings of female–male–male trios for which we have genetic data were fathered by the respective secondary male. Interestingly, this included two goslings fathered by a secondary male who was socially interested mainly in the primary male and not in the female. One reason may be that the secondary male unsuccessfully tried to mount the primary male, resulting in a copulation with the female instead. It is also feasible that this strategy may represent 'sneaking', where the bond between the males serves to keep the trio together throughout the breeding season. Unfortunately we have too few behavioural and genetic data to assess whether some males that ostensibly appear to be homosexual are, in fact, sneakers.

5.4 ALTERNATIVE REPRODUCTIVE STRATEGIES

Given that the majority of fledged offspring are produced by hetero-sexual pairs, it seems that this should be the preferred option, but establishing a heterosexual pair bond may not always be feasible due to factors such as limited partner availability or low individual quality. For such individuals, male–male bonds or polygamous bonds may represent a means of salvaging at least some reproductive success. Despite few direct fitness benefits, such alternative options may thus still be favoured, if they provide higher fitness than the other options would in a given context (Lyon & Eadie 2008). Furthermore, being heterosexually paired is no guarantee of reproductive success, as less than 20% of pairs ever manage to raise young to fledging and less than 7% do so repeatedly. Hence, even heterosexually paired individuals may benefit from other options (see below), particularly individuals of inferior individual quality (Lyon & Eadie 2008).

With the introduction of genetic testing techniques, it has become apparent that socially monogamous species are not necessarily sexually monogamous and that reproductive success may be quite different from what can be estimated from observation alone (see Griffith et al. 2002). Birds may gain (additional) reproductive success through extrapair fertilisation (EPF), intraspecific brood parasitism (IBP) or interspecific brood parasitism, although the latter is not an option in the Grünau greylag goose flock, where suitable hosts are not available. Hence, there are alternatives not only in the type of pair bonding, but also in reproductive strategies. These alternatives may be the means of gaining reproductive success for individuals engaged in alternative social bonds, and may also allow other individuals who do not raise young themselves to gain reproductive success, such as unpaired individuals or members of unsuccessful pairs. Indeed, the distinction between parasites with and without a nest of their own has been considered to be fundamentally important for understanding the evolution of intraspecific brood parasitism (Lyon & Eadie 2008). For example, nesting and non-nesting parasites have been shown to use different host selection criteria (Jaatinen et al. 2011). In addition, alternative reproductive strategies may be part of mixed strategies that could enhance the reproductive success of individuals that successfully raise young themselves (Åhlund & Andersson 2001). Alternative reproductive behaviours are common in the animal kingdom and have been described in a variety of goose species and other waterfowl, but the observed patterns are diverse and depend on the ecology of the species concerned (Anderholm *et al.* 2009b).

5.4.1 Extrapair fertilisations

More than 90% of all avian species are socially monogamous (Lack 1968), but true genetic monogamy occurs in only approximately 25% of these species (Griffith *et al.* 2002). The average frequency of extrapair paternity in socially monogamous bird species is 11.1% of offspring in 18.7% of broods, but may amount to more than half of the offspring in over 80% of the broods, as found in the reed bunting (*Emberiza schoeniclus*: Dixon *et al.* 1994). Rates of extrapair paternity are lower in species with a higher male contribution to parental care (Griffith *et al.* 2002) and, as such, it comes as no big surprise that the rates of extrapair paternity in geese are rather low. Rates of extrapair paternity were 2% in Ross's geese (*A. rossii*) and 5% in lesser snow geese (*A. caerulescens caerulescens*) despite 33% and 38% of forced extrapair copulations (Dunn *et al.* 1999), and Larsson *et al.* (1995) detected no extrapair paternity in 137 fledged young of barnacle geese. Similarly, we detected no extrapair fertilisations in a preliminary data set of 87 offspring from heterosexual pairs in the KLF flock. Hence, extrapair paternity does not seem to be a prevalent alternative reproductive strategy in our greylag geese. However, we were able to identify the fathers of six goslings from singleton mothers; two of these were fathered by homosexually paired males (see above) and four by heterosexually paired males. Consequently, at least 3.6% of all offspring analysed so far were fathered by heterosexually paired males outside their pair bond. Our preliminary data set, however, yielded no evidence for extrapair fertilisations among the offspring of heterosexually paired females. This pattern parallels findings by Choudhury *et al.* (1993), who detected no evidence for extrapair fertilisation in the females of nine barnacle goose families, but one potential case of extrapair fertilisation by one of the family males.

5.4.2 Intraspecific brood parasitism

Laying eggs in the nests of conspecifics enables individuals to gain the benefits of parental care without paying the often substantial costs.

Intraspecific brood parasitism (IBP) has been documented in over 200 species of birds and is particularly prevalent in some groups, including waterfowl (Lyon & Eadie 2008). IBP has been described as a means to flexibly adapt reproductive investment according to an individual's current condition and the environmental situation (Lyon & Eadie 2008). If prospects for success via nesting are low and/or costs of reproduction are high, females may employ strategies of low reproductive investment such as parasitic egg-laying without nesting themselves. On the other hand, when prospects are good, females may increase their reproductive success even above the maximum that could be gained in a single nest by adopting a mixed strategy of nesting and parasitic egg-laying. Furthermore, it may pay not to 'put all one's eggs into one basket', as harmful environmental conditions (e.g. floods, adverse temperatures), predation or compromised individual condition could lead to the failure of certain clutches, while others survive (Pöysä & Pesonen 2007).

IBP has been reported for 46% of the Anseriformes, but this may well be an underestimate (Geffen & Yom-Tov 2001). This high rate has been attributed to either the relatively large clutches of precocial birds compared with similarly sized altricial birds, to the start of incubation only after the ultimate egg is laid, or to the comparably small costs to hosts of rearing extra young that are precocial and able to sustain themselves shortly after hatching (Geffen & Yom-Tov 2001). In geese, IBP has been reported to result in 8–17% of extrapair young in 27% of barnacle goose families (Larsson *et al.* 1995; Anderholm *et al.* 2009a), and in approximately 6% of extrapair young in bar-headed geese (Weigmann & Lamprecht 1991). Lank *et al.* (1990) reported 7% of extrapair young while Dunn *et al.* (1999) detected only one case of IBP in another set of 80 young lesser snow geese. Our own preliminary data amounted to roughly 15% of the goslings being the consequence of IBP. Nests of unpaired females tended to be parasitised more often than those of paired females (75% vs 30% of nests; $\chi^2 = 3.606$, $p = 0.095$) and correspondingly, rates of extrapair young resulting from IBP were significantly higher in unpaired than in paired females (31% vs 8% of young; $\chi^2 = 5.883$, $p = 0.015$). This points to the importance of the presence of a partner for the female's success in guarding and defending her nest against parasites. Rates of IBP were also inversely correlated with dominance rank of the male (partial correlation, corrected for rank of the social mother, $r_p = -0.768$, $N = 10$, $p = 0.016$; Fig. 5.3), further corroborating the importance of the presence of a partner for nest defence. In contrast to these results, Weigmann and Lamprecht

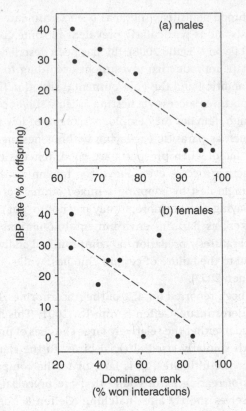

Figure 5.3   Relationship between dominance rank before the onset of breeding and rates of intraspecific brood parasitism (IBP) in ten greylag pairs.

(1991) found that nests of high-ranking bar-headed geese were parasitised more than those of low-ranking ones. Parasite–host dynamics may have differed in the two flocks because of differences in sex ratios and in the availability and spacing of nest boxes. Alternatively the opposite results may have been caused by methodological differences, as dominance ranks of the bar-headed geese were measured per pair and not, as in our flock, for each individual separately (see also Chapter 7).

To understand the strategies behind IBP one needs to know not only the rates and who is affected, but in particular, who the parasites are. Such data are hard to come by even in times when molecular methods are regularly applied in a wide range of species. The difficulty lies in having to sample not only the social parents and their offspring, but also a large proportion of the breeding population. Studies of waterfowl that have succeeded in identifying parasites have provided indications for IBP serving as an alternative reproductive tactic associated with lower fitness than nesting on one's own (Anderholm *et al.* 2009b), but they have also shown that some females may increase their reproductive success by laying parasitically *in addition to* nesting on their own (Åhlund & Andersson 2001; Reichart *et al.* 2010). In addition, several studies provide evidence that the host–parasite relationship may be influenced by kinship, i.e. that parasites are more closely related to their hosts than expected by chance (see Waldeck *et al.* 2008; Anderholm *et al.* 2009a; Jaatinen *et al.* 2009, 2011). Depending on relatedness, hosts may even show differential treatment of parasites during parasitic laying events (Åhlund 2005). Such inclusive fitness benefits may provide a functional explanation for the often lifelong maintenance of spatial proximity among female relatives in geese (see also Frigerio *et al.* 2001a; Black *et al.* 2007; Chapter 6). As we are still in the process of gathering genetic data from the flock, we have so far only succeeded in identifying five parasitic females. All five females were paired, all but one of those successfully hatched young in their own nest in the year of sampling, and interestingly, one of the five was a daughter laying into her mother's nest. Identifying the strategies of parasites and their relationships to their hosts in our flock is definitely a high priority for future investigations when genetic data become available in sufficient detail. The evidence available so far shows that genetic parents of fledged extrapair young were frequently individuals that successfully raised offspring themselves. In our flock, alternative reproductive strategies thus seem to enhance the reproductive success of already successful individuals, but may not increase the total proportion of successfully reproducing individuals.

SUMMARY

Greylag geese typically form heterosexual, monogamous pair bonds, but male–male pairs and trios (female–male–male or female–female–male) are also regularly observed. These alternative bonds are usually not specialist strategies but are formed flexibly over the course of individual life histories. In addition, greylag geese engage in a suite of alternative reproductive strategies, most notably intraspecific brood parasitism. These alternative social and reproductive strategies, however, seem to have only small fitness benefits except for those individuals that belong to the established, successfully breeding cohort. In particular, the engagement in male–male or polygamous pair bonds thus seems to represent a 'best of a bad job' strategy, which ensures social support and its associated benefits until better social and reproductive opportunities arise.

6

Beyond the pair bond: extended family bonds and female-centred clan formation

Family bonds extending from infancy into adolescence or even adulthood are a commonly described phenomenon in primate and non-primate mammals (e.g. red deer, *Cervus elaphus*: Albon *et al.* 1992; racoons, *Procyon lotor*: Gehrt & Fritzell 1998; spotted hyenas, *Crocuta crocuta*: Drea & Frank 2003). These associations often revolve around female kin. Primates and several cetaceans, for example, are well known for their matrilineal social organisation, where female off-spring remain with their mother for extended periods of time, or even for a whole lifetime. Via social support by kin, they may even inherit their mother's rank (Hinde 1983; Fedigan 1992). In African elephants (*Loxodonta africana*), the matriarchs are particularly important for the transfer of social knowledge (McComb *et al.* 2001). However, not only lineal kin (descendant offspring) are known to benefit from long-term associations with their family; collateral relatives (non-descendant kin) are also documented to stay together for extended periods of time. Certain defence behaviours, for example alarm calls, have been shown to benefit not only offspring but also non-descendant kin, for example siblings or cousins (Smith 1978; Schwagmeyer 1980; Cheney & Seyfarth 1981). Moreover, adult female white-faced capuchins (*Cebus capucinus*) are likely to form mother–daughter coalitions as well as coa-litions between full sisters and maternal, but not paternal, half-sisters (Perry *et al.* 2008).

In contrast to mammals, extended family bonds between lineal or collateral kin are rare in birds. Here, bonds are broken with the onset of the next breeding attempt at the latest (Zaias & Breitwisch

The Social Life of Greylag Geese: Patterns, Mechanisms and Evolutionary Function in an Avian Model System, ed. I. B. R. Scheiber *et al.* Published by Cambridge University Press. © Cambridge University Press 2013.

1989; With & Balda 1990; Verhulst & Hut 1996). In general, avian off-spring disperse as soon as they reach independence or shortly there-after (Nilsson 1990; Yoerg 1998). There are, however, exceptions to this rule. In some species, bonds between young and their parents last longer than usual. Here, the offspring stay to help raise consecu-tive broods (Brown 1987). Cooperative breeders are the best example (Langen 2000), but occasional helping behaviour is also known from basically non-cooperative species, including white-fronted geese (*A. albifrons*: Fox *et al.* 1995). Additionally, in barn swallows (*Hirundo rus tica*), predation risk is reduced not only through active mobbing by parents but also by siblings (Shields 1984).

Parent–offspring bonds extending beyond the offspring's inde-pendence are particularly uncommon in migratory birds (Wittenberger 1981). However, it has long been recognised that juvenile geese and swans stay with their parents for much of their first year ('primary fam-ily': Elder & Elder 1949; Prevett & MacInnes 1980; Scott 1980; Johnson & Raveling 1988; Lorenz 1988; Black & Owen 1989a) and commonly leave and return to the breeding grounds with them. Common explanations for those prolonged parent–offspring bonds include the guiding of off-spring during migration and transmitting foraging and social skills as well as the location of safe and productive foraging and roosting sites. Most juvenile geese leave the family unit during winter or spring (Prevett & MacInnes 1980; Johnson & Raveling 1988; Black & Owen 1989a), but a few studies report that bonds between parents and offspring, as well as between siblings, are maintained into a second year or even into adult-hood ('secondary family': Ely 1993; Warren *et al.* 1993).

From the offspring's perspective, individuals may benefit from a prolonged stay with their parents in terms of body condition. Five-month-old juvenile barnacle geese (*Branta leucopsis*) in family units were interrupted less during feeding than those that had already left the family unit. Furthermore, 10-month-old juvenile barnacle geese in family bonds were in better body condition after the first leg of migration than those without family bonds (Black *et al.* 2007). From the parents' perspective, associating with young through the spring enhanced the chances of returning with new offspring the following year. Both Lamprecht (1991) and Black *et al.* (2007) suggested that the offspring's contribution to detecting predators and fighting for forag-ing space enabled the parents to devote more time to their own forag-ing. This 'gosling helper hypothesis' is supported by data on family break-ups in barnacle and Canada geese (*B. canadensis*): parents toler-ated those juveniles that were most 'helpful' in agonistic encounters for the longest period of time in close proximity (Black *et al.* 2007).

Aggression of parents towards their young and parent–offspring conflicts seem to force the decision about whether and when to disperse (Trivers 1974). Similar to the observations reported by Black *et al.* (2007), data on the greylag goose (*A. anser*) flock at Grünau suggest that female offspring were the first to be chased away by their parents. While the timing of dispersal in barnacle goose juveniles might differ by weeks or even months, the dispersal of offspring at Grünau usually happens within a few days or weeks. This is in agreement with observations that juveniles of various *Anser* species generally remain with parents throughout the first year, while family break-up in *Branta* species is more variable (Owen 1980). In addition, life history data from the KLF flock show that young females were often either already paired or being courted at the age of 1 year. This may account for their earlier departure from the family unit. Hand-raised juveniles, on the other hand, when still accompanied by their human foster parent, also loosened the family bonds initially during the mating season, only to become stronger once again towards the summer (Weiß & Kotrschal 2004). This indicates that juveniles also play an active role when members of a family go their separate ways.

The advantages of goose family units staying together into spring are quite well understood, but knowledge of why family bonds extend past the first year is lacking. The occurrence of older offspring in secondary families was documented by Raveling (1969b), who reported that 15% of Canada goose yearlings rejoined their parents and that 2-year-old offspring also occasionally associated with their parents (D. G. Raveling, personal communication, in Ely 1993: p. 432). In the lesser snow goose (*A. caerulescens caerulescens*), adult offspring were not observed to do this, but 10% of the yearlings were still associated with their parents (Prevett & MacInnes 1980). The offspring of white-fronted geese were found to escort parents on average for 1.65 years (males) to 2 years (females) (Warren *et al.* 1993), sometimes extending to 3 years or more (Ely 1993; Kruckenberg 2005). Hence, the formation of extended family bonds is fairly common, but by no means the rule. In this chapter we therefore focus on questions pertaining to why individuals do or do not form secondary families or other extended family bonds in the first place, and ask the following questions:

- Are male or female offspring more likely to rejoin their parents?
- Are the individuals who join already pair bonded or are they still unpaired?
- Do parents accept all 'joiners' or do they reject some of their offspring?

- Are there multigenerational families as well?
- How often do 'tertiary families' (i.e. bonds with parents that extend into adulthood) form?
- Are there long-term associations between siblings too?

We also address the potential benefits of extended family bonds, which include offering social support. As we have devoted a whole chapter to this key element of greylag society (in Chapter 9), it is not discussed here. Instead, we ask whether there are long-term benefits of extended family bonds for future pair bonds and the reproductive success of parents and/or offspring.

6.1 LINEAL (PARENT–OFFSPRING) BONDS IN THE KLF GREYLAG GEESE

6.1.1 Secondary families

To address questions such as who joins parents in extended families, we surveyed the social relationships and reproductive parameters for the years 1992–2009. During that time, 87 families raised a total of 276 offspring to fledging. As in other goose species, these goslings stayed with their parents over their first winter; in other words, they remained in close proximity to the other family members, and showed socio-positive behaviours like greeting but little or no aggression towards one another. Instead, family members interfered in agonistic contests on each other's behalf (Scheiber *et al.* 2005a, 2009b; Chapter 9). Secondary families were less common than primary ones and were correlated positively with the number of primary families in the previous year (Fig. 6.1; Kendall's τ rank correlation: $\tau = 0.504$, $N = 18$, $p = 0.01$). The general appearance of secondary families was similar to that of primary families, although social support patterns differed (see Chapter 9). Of the 87 primary families observed between 1992 and 2009, 31 (35.6%) had formed secondary families one year later. We looked into the details of secondary family formation in a subset of 123 juveniles, fledged between 1998 and 2009, which survived their first year and whose parents also stayed alive. Of these, 34 (27.6%) joined their parents for a second year. Notably, in cases where parents fledged offspring in the following year ($N = 42$), only one (2.4%) of the juveniles from the previous year reattached herself to her parents. In contrast, 40.7% of the 81 juveniles whose parents did not produce fledged offspring the following year rejoined their parents to form a secondary family. Thus, the likelihood of joining the parents for a second year was significantly

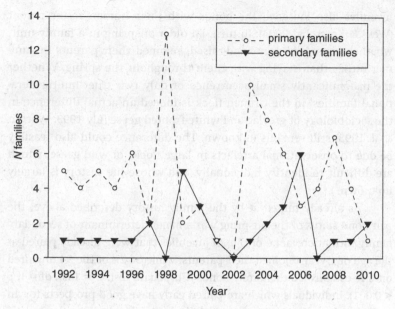

Figure 6.1 Number of primary (*white circles*) and secondary families (*black triangles*) from 1992 to 2010.

higher if the parents did not produce fledged offspring in that particular year (Chi-square test: $\chi^2 = 20.3$, df = 1, $p < 0.001$).

In fact, juveniles joining their parents when they had new offspring (i.e. multigenerational families) has occurred only twice in the history of the flock. In addition to the one case mentioned above, this involved four juveniles, hatched in 1993, who rejoined their parents in 1994 despite them having successfully raised another four young that year. The offspring from 1993 broke off in 1995, while the three females from 1994, but not the male, formed a 'regular' secondary family with their parents. Two unpaired females remained with their parents for a third and fourth year until the parents successfully raised young again in 1998. In white-fronted geese, new parental broods did not seem to be the reason for juvenile dispersal. Warren *et al.* (1993) observed at least one young of a previous brood being associated with parents and their new brood in four of five pairs. It is unclear, however, whether the older offspring remained associated with their parents throughout the complete breeding season or rejoined them after hatching or fledging of the new brood. In our flock of greylag geese, juveniles are chased away by their parents at the start of the breeding season without exception, and physically immature hand-raised juveniles also loosened the bonds with their human foster parent (Weiß & Kotrschal 2004). In turn, all older offspring in a family unit, whether goose-raised or hand-raised, rejoined their parents in summer rather than staying with them throughout the spring. Whether the insignificantly small occurrence of only two true 'multigenerational families' in the Grünau flock is indeed an actual difference in the sociobiology of greylag and white-fronted geese (Ely 1993; Warren *et al.* 1993) still remains unknown. The difference could also feasibly be due to observational artefacts in large flocks of wild geese, which are difficult to identify individually, and whose life history is largely unknown.

As already hinted at by the family history described above, the pair bond status of the offspring was a major determinant of secondary family formation; none of the 29 juveniles that were already paired at the age of 1 year rejoined their parents, while 36.2% of the 94 unpaired ones re-established the family bonds (Chi-square test: $\chi^2 = 14.5$, df = 1, $p < 0.001$). Individuals which are paired early have good prospects for an early first breeding attempt at a relatively small reduction in agonistic success, although pairs generally rank lower than secondary families. The limited availability of high-quality mates may constrain juvenile geese in forming pair bonds at a young age. Alternatively, they may attempt to rejoin the parents for another year or longer.

Furthermore, 'only children' reattached themselves to their parents more frequently than those who had siblings (Chi-square test: $\chi^2 = 4.99$, df = 1, $p = 0.025$). In spectacled parrotlets (*Forpus conspicillatus*), siblings form groups after dispersal from their parents, and offspring raised alone compensate for their lack of siblings by maintaining bonds with their parents for longer (Wanker 1999). Although siblings did not seem to have an influence in a previous analysis (Weiß *et al.* 2008), a reanalysis of the goose data with a larger sample size hinted at a similar phenomenon as that found in the spectacled parrotlets.

Effects of the offspring's sex on secondary family formation were variable; in some years, female offspring were considerably more likely to form secondary families with their parents, as for instance in the 1998 cohort. 1998 was a particularly successful year for the Grünau flock, with 44 fledged young in 13 families. In 1999, however, only two families raised young to fledging and, as a result, secondary family formation was high (55% of the 27 young that survived their first year rejoined their parents). Of these, 72.7% of the females but only 37.5% of the males reunited with their parents (Weiß 2000). Overall (1998–2009), however, the female bias that was apparent in some years was not evident, and males and females rejoined their parents at similar rates (35.7% of unpaired males and 36.8% of unpaired females). This is comparable to findings in white-fronted geese (Warren *et al.* 1993). Therefore, the payoffs of prolonged family life may generally be similar in the two sexes, but in some years environmental conditions such as food, mate and nest site availability may render extended family bonds particularly beneficial for females.

The most obvious advantage of reforming the family bond for a second year was an enhanced dominance rank when compared with the dominance rank of the same individuals while not in a family unit (Weiß *et al.* 2008). Also, juveniles at the age of 1 year that did not rejoin their parents ranked significantly lower than those that did (Weiß *et al.* 2008; Scheiber *et al.* 2009a). However, individuals won significantly fewer of their agonistic interactions when in a secondary family unit compared with the agonistic success of the same individuals while forming a primary family (Weiß *et al.* 2008). Data from hand-raised juveniles suggest that the reattachment to human foster parents is similarly important at one year after fledging. One-year-old hand-raised geese had agonistic success rates comparable to those after fledging when accompanied by their human allies. This effect declined towards winter despite the continuing presence of their human ally, and was no longer detectable at the next mating season, when the geese reached sexual maturity (Weiß & Kotrschal 2004). Furthermore, the presence of human allies enhanced the time spent feeding throughout the hand-raised geese' second year. Unlike agonistic success, the effect on feeding times was still noticeable at the end of the 2-year study period (Weiß & Kotrschal 2004).

Secondary family units provide similar, yet somewhat less pronounced, advantages as primary families. However, fewer than one-third of all young typically rejoin their parents for another year. Why is that? As we have seen earlier, none of the yearlings that had already formed a pair bond joined their parents for another year. Offspring who did not re-establish the family bonds formed pair bonds earlier than the ones that did (Weiß *et al.* 2008). In that way they gained a new social ally as well as an earlier chance of reproduction. In fact, the 'non-joiners' were younger than the 'joiners' when attempting to breed for the first time (Fig. 6.2; Mann–Whitney U test: $U = 70.5$, $N_{non\text{-}joiner} = 35$, $N_{joiner} = 8$, $p = 0.028$). At first glance, offspring that form secondary families thus have a reproductive disadvantage compared with their more independent peers. However, we did not detect any difference between joiners and non-joiners in the number of fledged

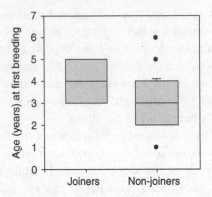

Figure 6.2   Age at first breeding of geese having rejoined ($N = 8$) or not rejoined ($N = 35$) their parents as yearlings to form a secondary family. *Boxplots* show medians and quartiles, *whiskers* 10th and 90th percentiles, and *circles* indicate outliers.

young produced per year over their lifetime ($U = 841$, $N_{\text{non-joiner}} = 61$, $N_{\text{joiner}} = 29$, $p = 0.573$), indicating that the benefits of secondary family life allow individuals to compensate for lost reproductive opportunities later in life. We may thus regard secondary families as a 'making the best of a bad situation' strategy, where individuals associate with former allies when reproduction failed or when no suitable mate is available. Prolonging the family bonds may allow parents to maintain their social status and body condition for the coming breeding season, while offspring may have a chance to gain quality as a mate and thus to increase the odds of securing a high-quality partner and successfully reproducing in the future.

Who decides on and initiates the formation of a secondary family? As both parents and offspring benefit from this aggregation, either would have a reason for making this arrangement. Indeed, it seems that both parties are involved in the decision, with the subadults taking the first step. Individual characteristics of the parents and offspring then influence the outcome: in a two-way process, subadult geese which intend to rejoin their parents initially seek their proximity. They also begin to utter 'vee'-calls again, a vocalisation that goslings direct to their parents and siblings in greeting. 'Vee'-calls are replaced by the adult version 'fluent cackling' when the goslings' voices break around the time of fledging (Lorenz 1991) and thereafter are only used when subadults try to re-affiliate with their parents. Parents then decide whether to tolerate the vicinity of their offspring or to displace them until the latter give up. The persistence of the young geese may play a role in this process. However, rejecting one offspring from the previous year does not necessarily imply that another will also be rejected subsequently, and neither does the acceptance of one youngster prevent others from joining the secondary family. In our flock, up to four young rejoined their parents simultaneously, although in the majority of cases only one or two subadults rejoin their parents. To date, we have never observed the opposite situation – parents actively seeking to coerce their offspring into joining them – and we are not aware of any studies investigating whether this is the case in other species of geese.

6.1.2 Tertiary families

Since 1998 we have observed seven offspring (i.e. two males, five females) from five different pairs maintaining tight bonds with their parents into adulthood. All of these individuals had already rejoined their parents in their second year. We term such bonds between parents and their adult offspring 'tertiary families' to distinguish them from bonds between parents and their subadult, non-dependent offspring ('secondary families'). Tertiary families break up in spring, just like the other family units, and in some cases reform once more in summer. The longest family bonds thereby lasted into the offspring's fourth year and we have not observed tight family bonds lasting beyond that point. Two of the females in a tertiary family were paired with a male in their third year, but still remained in a triumph ceremony group (for a definition see Chapter 11, p. 197) with their respective parents and one unpaired sister for one more year. The pair partners were occasionally 'left behind', but usually trailed the family. One of the females broke the bond with her parents after her sister's disappearance, but was often found within several metres of them thereafter.

Agonistic success of parents and offspring was similar in secondary and tertiary family units (Weiß et al. 2008), and the underlying mechanisms and functions of secondary and tertiary families also appear to be similar (see Chapter 9). The rare occurrence of tertiary families is probably due to offspring eventually forming pair bonds themselves, parents having new offspring, or the death of parent(s) and/or offspring. This suggests that ultimately there is more to gain by winning a mate or raising offspring than by remaining with the parents/adult offspring forever.

6.1.3 Even more prolonged parent–offspring associations?

To investigate whether family bonds eventually dissolve completely or whether less conspicuous associations are maintained between adult kin, we measured the distances between resting parents and their adult offspring ($N = 32$) and compared them with the distances between offspring and control individuals (I. B. R. Scheiber & B. M. Weiß, unpublished). As controls we chose five males and females close in age to the parents and neither related to nor paired with the focals or their parents. Resting individuals were identified from facial features and partially visible leg bands, and identities were confirmed when individuals got up and the leg bands were fully legible. During each observation, the distance of the focal to the father, mother and ten control individuals was measured with a laser distance meter. To obtain independent samples, we only observed each focal once during one resting period. The results of measuring distances between offspring and parents indicate that, in addition to offspring in tertiary family bonds, some offspring indeed remain significantly closer to their parents after family bonds had been 'officially' broken (Fig. 6.3; I. B. R. Scheiber & B. M. Weiß, unpublished). Distances between parents and their adult offspring, however, were quite large compared with the distances typically found in intact families (maximum distance up to 2 m). With two exceptions (<10 m), adult offspring were approximately 15 m away from their parents. This reduces the likelihood of active interference in agonistic interactions, which, as a matter of fact, was not observed at all during the study mentioned above. However, agonistic pressure may be lowered by reduced aggression among associated individuals. Notably, some individuals had preferred resting sites and the observed social preferences might therefore actually be caused by site fidelity. Individuals with or without a site preference, however, were no more or less likely to rest closer to their parents. Overall, these data show that associations between parents and offspring may persist well into adulthood, but these associations are loose and may represent an increased tolerance between kin rather than active social bonds.

6.2 COLLATERAL (SIBLING) BONDS AMONG ADULT KIN

Studies on lesser snow geese and white-fronted geese indicate that not only parents and offspring, but also siblings, may remain associated

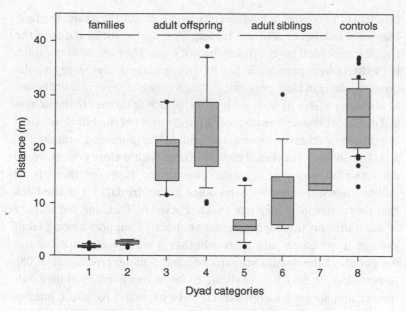

Figure 6.3 Intra-pair distances (m) between dyads of kin in various
social categories (1 = Parents–offspring dyads in primary families; 2 =
Parents–joiner offspring dyads in secondary families; 3 = Parents–adult
(>2 years) female offspring; 4 = Parents–adult male offspring; 5 = Adult
sisters; 6 = Adult sisters–brothers; 7 = Adult brothers; 8 = Random
controls of adult males and females). The data incorporated in this
graph have been gathered from various studies (Social category 1:
Scheiber *et al.* 2005a; 2 = Scheiber *et al.* 2009a; 3, 4 = I. B. R. Scheiber
& B. M. Weiß, unpublished; 5, 6, 7 = Frigerio *et al.* 2001a; 8 = I. B. R.
Scheiber & B. M. Weiß, unpublished). To be better able to compare data
and to have a common denominator, distances published previously
as categories are represented by giving actual values. *Boxplots* show
medians and quartiles, *whiskers* 10th and 90th percentiles, and *circles*
indicate outliers.

for extended periods of time (Prevett & MacInnes 1980; Ely 1993;
Warren *et al.* 1993). A number of observations, such as the provision-
ing of active support for a young female by the partner of her mother's
sister or the 'sharing' of partners by sister dyads suggest that sibling
associations also occur in our flock of greylag geese. Therefore, we
investigated bonds between adult siblings and measured the distances
between adult siblings and control individuals of the same age following a similar protocol as described above. This study showed that
adult females rested significantly closer to their sisters than to their

brothers or to control individuals (Frigerio *et al.* 2001a). Again, the findings were independent of site preferences, age or social status of the females; the oldest sister dyad in the flock was 12 years old at the time, both sisters were paired, one had offspring, and yet they were encountered regularly in close proximity to each other. Sisters typically rested at distances of 3–6 m from each other, which is further than the typical distance between members of a family unit (<2 m), but closer than the distances observed among adult lineal kin (approx. 15 m; see Fig. 6.3). The distance between sisters put them within close enough reach for active assistance in agonistic encounters; however, there is no indication in our extensive dominance hierarchy data set of the flock that sisters routinely help one another actively. This, and the scarcity of anecdotes on this topic, suggest that active support among adult siblings is extremely rare. Also, whether a sister was nearby or not did not affect the dominance rank of adult females (Weiß *et al.* 2008). Nonetheless, it may be beneficial to be surrounded by individuals from whom no serious aggression is to be expected. Agonistic interactions between adult sisters do occur but are rather rare and typically of low intensity. This suggests that sisters show an increased tolerance towards each other's presence, which may convey benefits such as better access to food sources, and 'incidental' social support through closer proximity to a sister's pair partner. Such tolerance may even lead to sisters 'sharing' a mate. Over the last 20 years, we observed the formation of 26 'trios' where a male was paired polygynously with two females (see Chapter 5). In 12 of these trios one or both females did not have any sisters, while in 9 cases the females were in fact sisters. Both sisters in a trio engaged in pair bond-related and sexual behaviours with the male, but other than performing the greeting ceremony, not with each other. Finally, sisters were occasionally observed to raise their young in close proximity and, in rare instances, even together (Weiß *et al.* 2008; Chapter 5).

Ely (1993) suggested that long-term sibling bonds might be very important in many aspects of the behavioural ecology of geese. Data from our flock support this idea for female siblings (Frigerio *et al.* 2001a). Again, the loose associations between female siblings seem to be characterised by increased tolerance. This, in turn, may facilitate the formation of stronger bonds when necessary, for instance to raise young together or to share a mate when another pair partner is not available. Similar aid-giving and co-breeding among related individuals is well known in species that are cooperative breeders (Brown 1987), but is known only anecdotally in geese. In contrast to females, males do not show similar distance patterns to either brothers or sisters (Frigerio *et al.* 2001a). However, males are known to sometimes form actual pair bonds with their brothers. These brother-pairs are often loud and aggressive and display the full range of pair bond-related behaviours, and in some cases also sexual behaviours (Kotrschal *et al.* 2006; Chapter 5). Brother-pairs thus differ profoundly from the associations found between adult sisters.

To conclude, in most vertebrates the young disperse from their natal site before, or at least when, they reach sexual maturity. Exceptions to this rule are multigenerational cooperative breeders (see Emlen 1997 for a review) and also the long-term lineal and collateral family bonds observed in waterfowl, which may persist for a lifetime (Black *et al.* 2007). Whereas in cooperative breeders family bonds are maintained because the young delay dispersal, continue to live with their parents, and may forego reproduction, extended family bonds in waterfowl, particularly geese, are broken during the breeding season but may reform afterwards. Females in particular seem to gain benefits from these associations (see also Chapter 9), which – in turn – may explain why female relatives often stay in close proximity to one another throughout life. In many aspects the extended family bonds, female-centred clans, and female philopatry in geese resemble mammalian for example cercopithecine primates (reviewed in Cheney & Seyfarth 2003), spotted hyenas (Drea & Frank 2003) or Coquerel's dwarf lemurs (*Mirza coquereli*, Kappeler *et al.* 2002), rather than other avian societies. Kappeler *et al.* (2002) suggested that the spatial clustering of female dwarf lemurs may represent an evolutionarily primitive state, which could then lead to the formation of permanent female bonded groups and matrilines in response to various other selective pressures.

SUMMARY

In this chapter we describe the various forms of extended lineal (parent–offspring) and collateral (sibling) family bonds in greylag geese. We focus on secondary families (rejoined offspring from earlier breeding attempts) and try to explain why they are formed. Male and female subadults join their parents at similar rates, but joiners are always single. Secondary families typically rank lower than primary families, yet parents achieve a higher rank than singletons or pairs without offspring. The emerging patterns suggest that secondary family formation is mostly a 'best of a bad job' tactic when no appropriate mate is available for subadults to pair with or if breeding conditions are unfavourable. This seems to be particularly true of females, for whom secondary families provide the means to stay in better condition than females with fewer or no social partners. Additionally, female siblings, as well as mother–daughter dyads, can be found in close proximity to one another when adult and already permanently paired. These bonds are loose and not comparable to primary and secondary goose families or matrilines in primates, but they structure a goose flock into female-centred clans. The extended family bonds, as well as the spatial proximity between close female relatives, indicate that females gain benefits through their maternal lineage. We have no indication that sons or brothers seek the proximity of their fathers/brothers in the same way that adult females do. This suggests that females are the driving force in structuring goose societies, while males leave their families in favour of their mates.

Part III   Costs and benefits of social life

7

Causes and consequences of aggressive behaviour and dominance rank

Aggressive behaviours can be advantageous in the acquisition and defence of food, mates and other resources (Ficken *et al.* 1990; Stahl *et al.* 2001; Vøllestad & Quinn 2003) and as such may profoundly influence survival and reproductive success (Kikkawa & Wilson 1983; Arcese & Smith 1985; Pusey *et al.* 1997). In fact, aggressive interactions do not just result from 'frustration', as some early psychologists held, but are an adaptive behaviour system (Lorenz 1963). However, high levels of aggression and achieving or maintaining a high dominance rank may incur considerable energetic costs and may increase the risk of injury or social stress (Hogstadt 1987; Kotrschal *et al.* 1998a; Sapolsky 2005). Aggressive behaviour is thus embedded in a tight trade-off between costs and benefits that will depend on an individual's own condition and life history as well as on the environment. A detailed understanding of the genetic and environmental bases of aggressive behaviour is necessary in order to understand which individuals will achieve and ultimately benefit from a high dominance rank (Richner 1989).

Aggressive behaviour and dominance rank are omnipresent in many animal societies, and geese are no exception. Particularly in a feeding context, aggressive interactions between geese are easy to observe. The vast majority of interactions are seemingly idle pecks and threats that probably serve to create space or act as 'reminders' of the status quo rather than establishing or changing dominance relationships (Fig. 7.1A). From time to time, however, individuals can be seen running towards and attacking others. Such attacks may escalate into prolonged chases on the ground or in the air during which the

The Social Life of Greylag Geese: Patterns, Mechanisms and Evolutionary Function in an Avian Model System, ed. I. B. R. Scheiber *et al.* Published by Cambridge University Press. © Cambridge University Press 2013.

assailant often keeps a firm grip on the target's tail or wings, leaving the winner with a beak full of feathers when the victim finally manages to break free (Fig. 7.1B). When matters become really serious, the opponents engage in a wing-shoulder fight, in which both opponents grab each other, usually at the shoulder, and beat each other using a horny knob on the wing front of the carpal area (Fig. 7.1C; Lorenz 1988). Usually these are of short duration, but – particularly during the mating season – they may last for minutes and will then leave both the winner and the loser completely exhausted, and sometimes even injured. Wing-shoulder fights are generally a male's business, but on very rare occasions we have also observed females dealing out wing blows towards other males or females, and even against a curious mute swan (*Cygnus olor*) or a European elk (*Alces alces*) which came too close to the female's freshly hatched goslings.

7.1 CAUSES OF AGGRESSIVE BEHAVIOUR AND DOMINANCE RANK IN GEESE

In geese and many other birds, males are more aggressive and higher-ranking than females (Arcese & Smith 1985; Kikkawa *et al.* 1986; Kotrschal *et al.* 1993; Poisbleau *et al.* 2006). Furthermore, it has long been recognised that the social status (being parental, paired without offspring, or unpaired) affects aggressive motivation and success: goose families are the highest-ranking social units in a flock, pairs are intermediate in rank, and unpaired individuals are lowest (Boyd 1953; Raveling 1970; Lamprecht 1986a; Gregoire & Ankney 1990; Poisbleau *et al.* 2006). Within families and pairs, however, agonistic interactions are extremely rare (Boyd 1953; Scheiber *et al.* 2009b). They are typically limited to the establishment of a dominance hierarchy within sibling groups in the first days after hatching (Kalas 1977) and to the time of family break-up, when parents start to go their own way as the next breeding season approaches (Black *et al.* 2007; Chapter 6). In addition to sex and social status, various other factors, such as family size, individual body size and age, have been found to affect aggressive interactions and dominance rank in some studies, but not in others (Mulder *et al.* 1995; Loonen *et al.* 1999; Stahl *et al.* 2001; Poisbleau *et al.* 2006). Furthermore, there is also experimental evidence for parental effects in juvenile barnacle geese (*Branta leucopsis*: Black & Owen 1987). Despite a considerable amount of literature on aggressive behaviour in geese, however, virtually nothing is known about the role of certain life history parameters such as pair bond duration, or about the

Figure 7.1   Agonistic interactions of different intensities. *Picture A* shows a typical threat posture (against an opponent who is not in the picture); *picture B* shows a flight chase where the attacker tries to grab the victim in mid-flight; *picture C* shows a wing-shoulder fight. © B. M. Weiß. A full-colour version is included in the colour plate section.

effects of season or the social environment. Indeed, the latter may be of profound importance for aggressive behaviour and dominance rank, which depend not only on an individual's own actions but also on the other individuals in the flock. Our understanding of aggressive behaviour may be further enhanced by not only knowing which parameters affect it, but also to what extent. Advances in statistical methods and an ever-growing long-term data set on aggressive interactions and dominance ranks in our goose flock allowed us to address these issues recently (Weiß *et al.* 2011).

7.1.1 Long-term data from the KLF greylag goose flock

The dominance relationships in our goose flock have been monitored regularly since the early 1990s. Data are systematically collected three times per year by the same observer (B. M. Weiß); in late summer after the flock has re-aggregated after moulting, in winter when social relationships in the flock are stable, and during the peak of the mating season in February. In each observation period, agonistic interactions are recorded for five consecutive days prior to and during the flock morning and afternoon feedings (see Chapter 1). In this situation, individual geese mostly attack and retreat independently of their pair partners or family members, while in other contexts pairs and families typically act as a single social unit that attacks or is being attacked together. For this and other reasons, some researchers have used the pair or family as the unit of analysis (Lamprecht 1991; Stahl *et al.* 2001). The feeding context at the KLF, however, allows us to measure aggression rates and dominance rank at an individual level, and thereby to relate these traits also to the individuals' characteristics. Hence, the observational data on agonistic interactions are used to calculate each *individual*'s aggression rate, interaction rate and dominance rank in the respective period. Dominance rank in geese is mostly, but not strictly, linear (i.e. if individual (or unit) A is dominant over B, and B is dominant over C, C could still be dominant over A). We thus do not assign a strict hierarchical order to all the individuals present during the observation period, but rather calculate dominance rank as the percentage of defeated individuals out of the total number of opponents for all individuals observed interacting at least five times (Weiß *et al.* 2011).

To look beyond the obvious effects of sex and social status, we analysed repeated measurements of aggression rates and dominance rank of 445 individuals using generalised linear mixed models (GLMMs). Mixed models provide a powerful tool for modelling a large

number of effects such as age, sex, social status or season simultaneously, thereby accounting for the presence of other effects. They allow
the incorporation of repeated measures on the same individuals, so
that we can make full use of the data rather than having to calculate
means across individuals. Furthermore, the importance of the various effects can be compared by means of the calculated effect sizes
(Garamszegi *et al.* 2009), which allows a more accurate interpretation
of the overall picture than significances alone. As individual needs, and
accordingly the costs and benefits of aggression, may change throughout life, we performed the analysis not only on the overall data set, but
also separately for the main life history stages (juveniles, subadults,
unpaired adults, paired adults and parents) to assess whether and how
the determinants of aggressive behaviour change throughout an individual's life.

7.1.2 Individual careers

Before becoming engrossed in the hard data from the flock, two anecdotes will illustrate some of the individual careers in dominance rank
that we have observed over the years. Although each career follows
a different path, in most cases we can readily explain why a particular individual rose or fell in rank at any given time. While the exact
stories are diverse, the causes are frequently very similar, and these
anecdotes will help to illustrate the complexities that cause inter- and
intra-individual variation in aggressive behaviour and dominance
rank.

Punki was a male hand-raised in 1992. He was a high-ranking
youngster when still accompanied by his siblings, dropped somewhat
in rank when the sibling bond dissolved in the following year, but
was still intermediate in rank (Fig. 7.2a). He rose again in rank after
forming a pair bond with his brother *Pirat*, but *Pirat* disappeared a few
months later, leaving *Punki* unpaired once more. *Punki* subsequently
formed a pair bond with a female of the same age, *Diana*, and maintained a fairly high-ranking position in the following years. In 1998
the pair fledged two offspring. The latter remained with their parents
for nearly 2 years, resulting in *Punki*'s high rank in 1998 and 1999.
The pair was not successful in raising young to fledging in the following years and resumed a rank position similar to that before their
reproductive success. In the spring of 2002, *Diana* and her 1-day-old
goslings were taken by a fox, and *Punki* dropped to the very bottom
of the flock hierarchy. In the following winter, *Punki* was courted by

another male and formed a rather loose pair bond with him that did not boost *Punki*'s rank for about 2 years. When the bond had gained some stability, *Punki* rose again in rank but, until his death in 2011, never managed to regain the relatively high dominance rank he had held in earlier years.

Timber is a female who was raised by her parents in 1999 in an extraordinarily large sibling group of 12 goslings. The super-sized family was by far the highest-ranking social unit in the flock. Fresh out of her family unit, *Timber* formed a pair bond with the same-aged, hand-raised male *Murphy* at the young age of 11 months. The pair bond has been stable ever since, with *Murphy* quickly becoming a high-ranking male. Throughout the years, *Timber* has been rather low-ranking, with the exception of the two years when *Timber* and *Murphy* successfully raised young to fledging (Fig. 7.2b).

7.1.3 Life history effects on aggression rates and dominance rank

Punki's and other examples of individual careers (Fig. 7.2) nicely illustrate the effects of social status and sex on dominance rank which were already recognised in the early years of goose research. The mixed model analyses confirmed that social status and sex had large effects on aggression rates and dominance rank in geese that remained evident throughout all the life history stages (Weiß *et al.* 2011), although there were other, even larger, effects (Fig. 7.3).

In addition to these rather evident effects, several other factors seem to be in the mix (Fig. 7.3; see also Chapter 3). For instance, age has been shown to influence aggressive behaviour in a number of species (Arcese & Smith 1985; Estevez *et al.* 2003), but studies considering the effects of age on goose aggressive interactions and dominance relationships produced equivocal results (Lamprecht 1986a; Black & Owen 1987; Stahl *et al.* 2001). We did detect an overall positive relationship between age and aggression rates as well as dominance rank, but the very small effect sizes in all cases suggest that age *per se* plays only a minor role in aggressive behaviour of geese.

At first glance, the raising history of an individual, i.e. whether it was hand-raised or goose-raised (see Chapter 1), has no effect on aggression rates and only a small effect on dominance rank. This picture is quite different, however, when looking at the different age classes. In juvenile geese, i.e. those up to 1 year of age, goose-raised individuals are frequently more aggressive and more dominant than

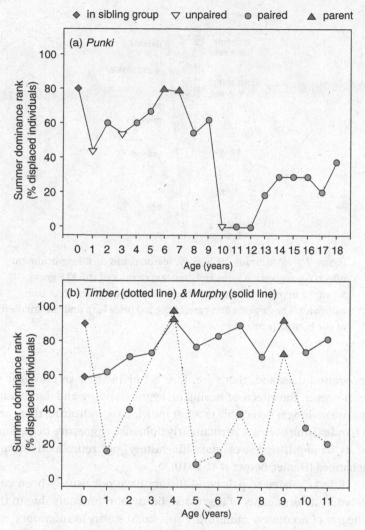

Figure 7.2   Lifetime summer dominance rank of (a) the male *Punki* and (b) the pair *Timber* (*dotted line*) and *Murphy* (*solid line*).

hand-raised ones (Weiß *et al.* 2011). This is very likely to be due to the social support that goose-raised individuals receive from their parents, while many of the hand-raised geese are no longer accompanied by their human foster parents and thus only have their siblings as social allies (see Chapter 9). In subadults, however, the opposite is the case – possibly, hand-raised geese have had to learn to manage their relationships in the flock without parental support at an earlier stage than

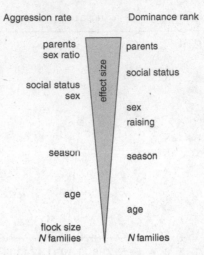

Figure 7.3 Significant life history, seasonal and social environment effects on aggression rates and dominance rank of the KLF geese, arranged in order of their importance as derived from effect size estimates. The largest effects are at the top (*wide bar*), and the smallest at the bottom (*narrow bar*).

goose-raised ones and, therefore, fare better in their second year of life. However, the effects of raising on aggression rate and dominance rank are no longer detectable in adult individuals, indicating that careful hand-raising does not permanently influence aggressive behaviour, just as hand-raising leaves most life history and reproductive traits unchanged (Hemetsberger *et al.* 2010).

Effects related to pair bond history have not usually been considered as determinants of aggressive behaviour, probably due to the difficulty of accurately monitoring pair bond status in migratory and only partially marked populations. The detailed monitoring of social relationships that ranges back to, and even beyond, the establishment of our flock here at Grünau (Chapter 1) has allowed us to also consider such parameters as pair bond type, pair bond duration, the partner's rank, and previous reproductive success as possible determinants of aggression rates and dominance rank. Thereby, the pair bond type, i.e. whether an individual was engaged in a heterosexual pair bond, a homosexual pair bond, or a trio (a male with two females or a female with two males; see Chapters 2 and 5) turned out to be a fairly strong determinant both of aggression rates and of dominance rank of adult paired geese. As males are typically more aggressive than females, and three individuals intuitively seem stronger than two, we might expect male–male pairs and trios to be more aggressive and higher-ranking than heterosexual pairs. However, why are heterosexual pairs then the 'default' bond in geese (Chapters 4 and 5) if they have a competitive disadvantage compared with other pair bond types? In fact, this is not the case at all. On the contrary: heterosexual pairs are generally more aggressive and win against more of their opponents than trios, who take the intermediate position. Homosexual males are actually the least aggressive and dominant of the bonded individuals (Weiß et al. 2011). These results are in contrast to reports of homosexual ganders' behaviour, which has been described as overtly aggressive in the past (Huber & Martys 1993); this may have been an effect of the general 'loudness' and behavioural conspicuousness of such pairs. A fairly recent, detailed study of homosexual pair bonding in geese showed that only a few individuals match the earlier descriptions of homosexual males, while the majority seem to engage in 'homosocial' bonds as a 'best of a bad job' tactic when female partners are not available (Kotrschal et al. 2006; Chapter 5). The prevalence of 'homosocial' bonds among the male–male pairs may thus explain why male–male pairs generally do not dominate heterosexual pairs. A similar explanation may be appropriate for the lower aggressiveness and rank of trios; like male–male pairs, trios seem to be 'a best of a bad job' strategy, at least for some of their members (Chapter 5). This may affect aggressive motivation and investment as well as coordination between the members of a trio, particularly in dense flock situations, where keeping track of the position(s) of one's social allies is presumably more difficult. In addition to the pair bond type, aggression rates and dominance ranks generally increase with increasing pair bond duration and the partner's rank. However, comparable to age effects, effect sizes of these parameters were very small and thus belong to the list of parameters that have statistically significant, yet only minor, effects on the aggressive behaviour of geese. Whether or not a pair had previously managed to raise young to fledging had no effect on their present aggressive behaviour at all (Weiß et al. 2011).

Several studies of geese have reported that large families dominate small ones in agonistic interactions (Boyd 1953; Gregoire & Ankney 1990; Loonen et al. 1999), but this need not always be the case (Lamprecht 1986a; Mulder et al. 1995; Poisbleau et al. 2006). Loonen et al. (1999) suggested that family size may be important only when food is scarce or patchily distributed and competition is therefore high. Along these lines, we would not expect effects of family size in our flock, where food is available almost *ad libitum*, yet we found family size to have a significant positive effect on aggression rates and dominance rank. A closer look at the mixed model statistics, however, showed that once again the importance of this effect was very low (Weiß *et al.* 2011). Consequently, our results fit well with Loonen *et al.*'s (1999) suggestion that competition drives the importance of family size for agonistic success. Importantly, this example also illustrates that looking only at statistical significance, but not at effect size, may result in a rather distorted version of the overall picture.

In contrast to family size, family type had significant and pronounced effects on aggressive behaviour. Primary families, i.e. parents and their juvenile offspring, were more aggressive and achieved a higher dominance rank than parents accompanied by their subadult offspring (secondary families), indicating that parents invest more into aggressive behaviour when their offspring are younger and more dependent on parental guidance and support (see also Chapter 9).

7.1.4 Environmental variation

Our flock is non-migratory and data on the dominance structure of the flock were collected under fairly standardised conditions (same location, same relative time of day, all or most of the flock present). Environmental variation may thus have a smaller impact on this flock's aggressive behaviour than in other flocks, but one key variable is nonetheless seasonal variation. Seasons may affect behaviour via changes in daylight, temperature and food abundance, but also through changes in internal states (e.g. hormones) and in individual needs that vary throughout the year (Wilson & Boelkins 1970; Hill 1997). In late winter and spring, females, for instance, will have to maximise energy intake in preparation for the breeding season. Males, on the other hand, will need to make this possible for their female partner by providing access to the largest and/or best food sources, or will need to engage in fights to obtain or defend a partner. In summer, when the flock re-aggregates after moulting, dominance relationships may need to be re-confirmed or re-established, particularly if individuals have changed or lost a partner in the previous breeding season. In winter, on the other hand, saving energy in cold temperatures may be a prime driver of behaviour. Astonishingly, however, seasonal variation and its effects on aggressive behaviour have received little attention. One exception is a study by Tarvin and Woolfenden (1997), who showed that female blue jays (*Cyanocitta cristata*) at a feeder were more aggressive than expected prior to the breeding season. Males were rather less aggressive throughout the winter months and early spring, but were more aggressive than expected in late spring. The authors suggested that the increased aggression of females was due to the higher energy demands in preparation for breeding and that of males due to feeding of the nestlings. Furthermore, changes in aggressiveness had no effect on the jays' dominance status.

Compared with other effects, the seasonal effects detected in our recent study were of rather moderate size, but they affected all life history stages and showed the biggest differences between them (Weiß et al. 2011). The observed patterns thereby neatly reflect the changing demands of individuals over the course of the year and life, but also suggest an optimisation of behavioural investment according to the agonistic pressure exerted by other flock members. For instance, individuals needing to secure paternity and essential resources for breeding are particularly aggressive and rank high during the mating season, while those who are too young or unaffiliated lie low during that time (Fig. 7.4). Parents achieve the highest dominance rank in summer when their offspring need to establish their place in the flock for the first time, thereby probably boosting the juveniles' position in the hierarchy as well as their survival in the first winter. Parental agonistic investment appears to decline over the course of the year and results in the parents' lowest agonistic success during the mating season, when family bonds are about to break up and parental needs come to the fore. It may seem counterproductive for parents to reduce their agonistic investment at a time when resources for breeding need to be secured; however, despite their seasonal low, parents still dominate the vast majority of their conspecifics during that time (Fig. 7.4). The female thus continues to have access to essential resources, but in addition may improve her energy balance by directing less effort into defending her offspring. The higher dominance rank of juveniles and subadults during winter may be a result of the relatively low agonistic pressure by other high-ranking flock members during that season. Also, juveniles may still have to gather social knowledge as to whom they can or cannot safely displace, and how this is dependent upon one's distance to social allies. This could be another reason why juvenile geese do not rank highest in summer, when their parents do.

7.1.5 Effects of flock structure

The life history variation described in the previous paragraph provides some indications that aggressive behaviour may depend not only on one's own motivation and state, but also on that of others. Effects of

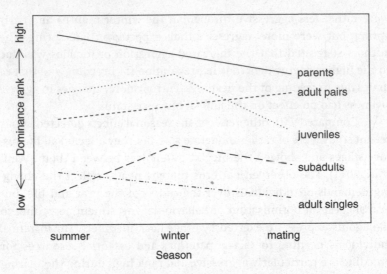

Figure 7.4:   Schematic illustration of the seasonal changes in dominance rank for the main life history stages.

the social environment on aggressive behaviour and dominance relationships have received little attention in birds, but some studies do provide evidence for such effects. In domestic fowl (*Gallus gallus domesticus*), increasing group size was accompanied by decreased aggression between group members (Estevez *et al.* 1997, 2003), and the dominance rank of red grouse (*Lagopus lagopus scoticus*) and Japanese quail (*Coturnix coturnix japonica*) was found to be influenced by the number of same-sex brood mates (Boag & Alway 1980; but see Arcese & Smith 1985 on song sparrows, *Melospiza melodia*). In our recent mixed model analyses, we included several descriptors of flock structure, namely flock size, sex ratio and, due to their high rank position, the number of families in the flock. While we expected some effects of flock structure, it came as quite a surprise that the sex ratio of the flock not only had a significant effect, but that it turned out to be the effect with the largest effect size of all the possible determinants considered (Weiß *et al.* 2011). This was true for aggression rates in general, but was even more obvious when looking at the different life history stages separately. Subadult geese, in which effects of sex ratio were absent, were the only exception. In the remainder (juveniles, adult singles, pairs and parents), individuals were more aggressive when the flock contained more males, independent of their own sex. This could indicate a bias towards attacking males, or a higher agonistic pressure due to the higher number of males which, in turn, results in individuals also initiating more agonistic interactions themselves. Such a redirection of attacks against other flock members may allow individuals to avoid the negative effects of a retreat by reducing tension, reducing the rate of renewed attacks, or restoring access to resources (Aureli *et al.* 1993; Chapter 9).

The sex ratio of the flock was also the strongest determinant of dominance rank in juveniles and subadults, but had no significant effects on adult dominance rank. As would be expected, subadults had less success in agonistic interactions at times when the flock contained more males and agonistic pressure was presumably higher but, interestingly, the reverse was true for juveniles. As suggested earlier, it might be particularly important for juveniles to demonstrate the family's high rank to other high-ranking individuals (Scheiber *et al.* 2009b; Chapter 9), and males may therefore be preferred targets, while juveniles still benefit from parental support. Arcese and Smith (1985) proposed that the accumulated experience of an individual in agonistic interactions is a major determinant of dominance. In our stable and fairly small flock, the sex ratio very probably no longer influenced adult dominance rank, as individuals may have learned their own position in the dominance hierarchy as well as that of the other flock members. Quite contrary to the effects of sex ratio, increasing flock size and the number of families did reduce aggression rates, but – like age and most pair bond-related effects – these effects were transient and small.

7.1.6 Parental effects

Parental effects comprise any influences the parents might have on their offspring's behaviour and thus might include effects related to the actual presence of the parents while family bonds persist (e.g. social support; see Chapter 9), long-lasting consequences of non-genetic effects (e.g. hormonal composition or size of the eggs, parenting styles; see Chapter 3), or a heritable genetic component. In our flock, the identity of the parents affected an individual's dominance rank more than any of the other effects and, next to the sex ratio of the flock, was the second main predictor of individual aggression rates (Fig. 7.3). Parental effects were evident in individuals of all ages, but were absent (or not significant) in unpaired adults and parental birds (Weiß *et al.* 2011). These two social classes are positioned at the very bottom and top of the dominance hierarchy, respectively, and accordingly may be subjected to factors like social stress or fierce competition more than any other individuals. At these positions in the flock hierarchy, the social environment may thus suppress the expression of parental effects.

Parental effects were stronger in juveniles than in other life history stages, and as juveniles typically still form tight bonds with their parents, this points to some role of the parents' actual presence. However, the parents' current dominance rank contributed very little to their offspring's success (Weiß *et al.* 2011). Parents, however, might also differ in the maintenance of spatial proximity to their offspring or by providing different amounts of social support, thereby influencing the frequency and success of their offspring's agonistic interactions. However, effects related to the actual presence of parents cannot fully account for the observed parental effects, as the latter were also present in adult birds, long after the disintegration of family units. Also, a cross-fostering study in barnacle geese (Black & Owen 1987) showed that, in the absence of their parents, juvenile geese from dominant parents ranked highest, young from subordinate parents raised by dominant parents ranked intermediate, and those from subordinate parents ranked lowest, suggesting also that the 'indirect' parental environment and genetics contribute to rank acquisition in juvenile geese.

To disentangle the different forms of parental effects and, in particular, the role of an individual's genetic predisposition, we employed recent statistical advances for studying quantitative genetics in natural animal populations. The so-called 'animal model' is a form of mixed model analysis that uses multigenerational pedigrees combined with phenotypic records (Kruuk 2004). This allows the separation of additive genetic effects from other sources of variance, such as maternal or parental effects, common environment (e.g. being nestmates) or permanent environment effects (i.e. persistent differences between individuals not attributable to additive genetic variance). These analyses revealed that heritabilities (i.e. the percentage of total phenotypic variance explained by additive genetic effects) and non-genetic parental effects were generally small (Weiß & Foerster, 2013). Permanent environment effects were slightly larger than the other components, pointing to a personality component in aggressive behaviour, as described in detail in Chapter 3.

This overall picture differed between the two main life history stages – i.e. juveniles and paired adults (Weiß & Foerster 2013). In juveniles, heritability estimates for aggression rates did not differ much from those of paired adults, but those for dominance rank were about twice as high as in paired adults. Permanent environment effects were basically absent in juveniles, while non-genetic parental effects accounted for a substantial part of the phenotypic variation in aggression rates and dominance rank. In adult pairs, additive genetic and parental effects were small or absent, but permanent environment effects were pronounced. Furthermore, the identity of the partner also explained a fair amount of the observed variation in aggression rates and dominance rank of adult pairs (Weiß & Foerster 2013). These patterns suggest that the observed parental effects are composed of a large direct component related to the actual presence of the parents. Furthermore, these results indicate that parental support and genetic predisposition may be pivotal components for establishing a juvenile's position in the social hierarchy. In adult geese, on the other hand, individual experience and the social environment, for instance having a suitable partner, contribute more to agonistic success than genetic or parental components.

Different heritabilities in different individuals, such as the age-dependence mentioned above, may be indicative of different genetic optima in varying spatial or temporal environments, or in different individuals. Such evolutionary constraints may provide an explanation for one of the most puzzling questions in evolutionary biology, namely the maintenance of heritable genetic variation in natural populations, while selection should continually favour certain genotypes at the expense of others (Fisher 1958). Sex differences represent another potential source of evolutionary constraints, and sex-dependent heritabilities have been reported for a variety of traits in humans and non-human animals (Nol *et al.* 1996; Galsworthy *et al.* 2000; Jensen *et al.* 2003). Our study provided no evidence for sex differences in heritabilities *per se*, but the comparison of adult male and female heritabilities revealed a rather intriguing detail; namely that heritability estimates of dominance rank were considerably higher when estimated separately for males and females than when estimated across the two sexes (i.e. with both sexes in the same data set). This may indicate that genetic effects on dominance rank are mainly passed on from mother to daughter and from father to son, but not from parents of one sex to offspring of the other. This impression was supported by a genetic correlation, which describes the proportion of variance that two traits share due to genetic causes. The genetic correlation between male and female dominance rank was significantly smaller than 1 (Weiß & Foerster 2013), which could arise if different constellations of genes affect a trait (Falconer 1952). In contrast, heritability estimates for aggression rates did not differ much between the sex-specific and the overall adult data set and showed a high genetic correlation between the sexes.

Our results thus suggest that dominance rank, but not aggression *per se*, is regulated by different genes or differential gene expression in adult males and females. Also, sex-specific constraints imposed by the environment may influence the expression of dominance rank through mechanisms such as hormonal influences on gene expression and regulation or other non-genetic factors that are correlated with sex. Goose parents thus influence their offspring's aggressive behaviour through genetic and non-genetic components that act in a complex, age- and sex-dependent manner.

7.2 CONSEQUENCES OF AGGRESSIVE BEHAVIOUR AND DOMINANCE RANK

By applying advanced statistical methods to long-term data on aggressive behaviour from our goose flock, we were able to show that agonistic behaviour in geese is a product of parental effects, including a heritable component, and a complex interplay of personality (Chapter 3), an individual's life history stage, and the social environment. The heritable component thereby represents a key ingredient for evolutionary changes in dominance and aggression. However, the other key ingredient is the selective advantage of being dominant or aggressive through increased survival and reproduction. Evidence for such an advantage comes from a wide range of taxa and suggests that, while the causes of a high dominance rank and aggression are highly diverse in different species, the benefits are rather universal; dominance and related behaviours have been shown to increase access to limited resources (Ficken *et al.* 1990; Prop & Deerenberg 1991; Stahl *et al.* 2001) and thereby enhance survival (Kikkawa 1980; Arcese & Smith 1985; Stahl *et al.* 2001) and reproductive success (Dunbar 1980; Kikkawa & Wilson 1983; Black & Owen 1987; Pusey *et al.* 1997). On the flip side of the coin, however, too much aggression may have a range of deleterious effects (Hogstadt 1987; Kotrschal *et al.* 1998a; Verhulst & Salomons 2004; Sapolsky 2005) that may counteract the benefits; therefore a net benefit of being dominant and/or aggressive cannot be taken for granted.

In geese, studies certainly indicate the benefits to dominants in obtaining access to resources. Dominant barnacle geese occupy favourable food patches that were detected by subdominants (Stahl *et al.* 2001), and analyses of brent geese (*B. bernicla*) faeces showed that dominant individuals obtained more high-quality plants than low-ranking ones (Prop & Deerenberg 1991). Furthermore, dominant female bar-headed geese (*Anser indicus*) had higher food pecking rates than subordinate ones (Lamprecht 1986b). Effects of dominance rank on survival are suggested in a study of barnacle geese by Stahl *et al.* (2001), who found that those females that returned to the breeding area in a given year had been significantly more dominant in the previous year than those that did not return. However, as the exact fate of the non-returning birds remains unknown, this result could also be explained by other reasons, such as subordinates shifting their breeding location to areas with less competition. Also, barnacle goose pairs associating with their offspring into the following spring were more likely to return to the wintering grounds with fledged offspring in the following year. This has been attributed to the continued high rank that parents can maintain with the help of their goslings ('gosling helper hypothesis': Black *et al.* 2007).

More systematic evidence for the fitness consequences of aggression comes from another study on barnacle geese; non-parental birds that successfully reproduced in the following year were involved in significantly more conflicts (wins and losses) than those that did not (Black & Owen 1989b). Whether these results also extend to success in agonistic interactions (i.e. to a high dominance rank) was not reported. The only comprehensive analyses of the effects of dominance rank on reproductive success thus seem to be the studies of Lamprecht (1985, 1986a, 1986b) in a flock of captive bar-headed geese. He showed that winter dominance correlated positively with the number of fledged offspring in the subsequent breeding season. Specifically, this correlation seemed to come about due to winter dominance influencing the likelihood of breeding and hatching success, but was not due to differences in the rearing success of dominant and subordinate pairs.

Based on the vast body of data available from our flock, we have recently begun to explore the role of dominance and aggression for individual fitness in more detail. As mortality is highest in juveniles and subadults in many birds (Loery *et al.* 1987; Martin 1995), we focused on individual survival in the first two years of life. Of the almost 300 geese that fledged between 1995 and 2009, those who died before the age of 2 years were significantly less aggressive and had a lower dominance rank in their first summer than those that survived and recruited into the breeding population (Fig. 7.5; Mann–Whitney *U* test,

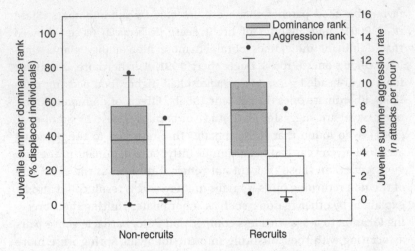

Figure 7.5   Juvenile summer dominance rank (*grey*) and aggression rates (*white*) of individuals recruiting and not recruiting into the breeding population. *Boxplots* show medians and quartiles, *whiskers* 10th and 90th percentiles, and *circles* 5th and 95th percentiles.

dominance rank: $Z = -2.980$, $N_{died} = 45$, $N_{survived} = 198$, $p = 0.003$; aggression rate: $Z = -3.171$, $N_{died} = 47$, $N_{survived} = 204$, $p = 0.002$). Furthermore, juvenile aggression rates and dominance rank also correlated negatively with female age at first pair bond (dominance rank: $r_s = -0.235$, $N = 89$, $p = 0.026$; aggression rate: $r_s = -0.229$, $N = 94$, $p = 0.026$). Hence, the more aggressive and dominant females paired at an earlier age than the less aggressive and dominant ones, thereby having earlier opportunities for successful reproduction. It must be noted, however, that these effects were not pronounced, as the correlation coefficients were rather small. There was no significant relationship between juvenile aggressive behaviour and age at first pair bond in males (dominance rank: $r_s = -0.034$, $N = 94$, $p = 0.743$; aggression rate: $r_s = -0.089$, $N = 94$, $p = 0.391$).

For adult geese, we explored the relationship between subsequent reproductive success and dominance-related behaviour in the preceding year. For this purpose we performed mixed model analyses on male and female reproductive success from 1996 to 2009, correcting for repeated measures of the same individuals and yearly variation. The effects of dominance rank and aggression rates during the mating season on the production of fledged young in the subsequent breeding season were modelled simultaneously. Results obtained with aggression rates and dominance rank during the preceding winter were similar to those of the mating season and are not addressed further. Both males and females who were successful parents in a given year had had a higher dominance rank in the preceding mating season than individuals of the same sex that did not raise offspring to fledging (Fig. 7.6). Notably, these effects were significant but small (GLMM, males: Wald = 10.77, df = 1, 409.4, p = 0.001; effect size = 0.031 ± 0.009; females: Wald = 11.17, df = 1, 364.7, p < 0.001; effect size = 0.029 ± 0.009). Accounting for the effects of dominance rank, aggression rates in males did not further explain variation in reproductive success (GLMM, Wald = 0.36, df = 1, 409.4, p = 0.547). In females, successful mothers had, in fact, had lower interaction rates (including aggression rates) in the preceding winter than unsuccessful ones (GLMM, interaction rate: Wald = 4.47, df = 1, 296.3, p = 0.035; effect size = −0.17 ± 0.08; aggression rate: Wald = 5.83, df = 1, 337.8, p = 0.016; effect size = −0.406 ± 0.168). Successful females thus probably had better access to resources due to their higher dominance rank, but spent less energy and time on achieving this. This may have resulted in a better energy balance of subsequently successful compared with unsuccessful females, despite the supplemental food provided for our flock. Within the group of successful parents, neither aggression rates nor dominance rank during the mating season had a significant influence on the number of subsequently produced offspring (GLMM, all p >0.05).

The data reported above fit well into the overall picture of the benefits of dominance and aggression in animals, but do not include an assessment of the potential costs. Evidence for such costs comes from a study comparing stress hormone metabolite levels determined from goose faeces in males of different social status throughout the annual cycle (Kotrschal *et al.* 1998a). In particular, parental males had higher corticosterone metabolite levels than paired or unpaired males for more than half the year. This may well relate to the need for maintaining a high dominance rank in this social class. Furthermore, anecdotal evidence portrays a range of deleterious effects of aggression, from failing reproduction to the death of an individual. We observed the latter only once, in the instance of *Stromer*, a 13-year old male, who got involved in one of spring's regular 'soap operas'. The centre of attention was the female *Tiger*, the flock's 'femme fatale'. She was courted regularly by a suite of adorers each and every spring, irrespective of her pair bond status. In the process, *Stromer* was engaged in a wing-shoulder fight with *Tiger*'s most promising aspirant. The fight lasted for several minutes and by the end the fully exhausted

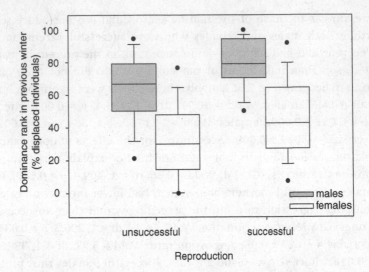

Figure 7.6   Dominance rank of successfully and unsuccessfully reproducing male (*grey*) and female (*white*) greylag geese in the preceding mating season. *Boxplots* show medians and quartiles, *whiskers* 10th and 90th percentiles, and *circles* 5th and 95th percentiles.

opponents were clinging on to each other only with their beaks. As a result of this, *Stromer*'s lower beak was torn off, leaving him with a fatal injury. To our knowledge this is the only reported casualty directly resulting from a fight between ganders.

Similarly, reproductive failure is often the consequence of a male trying to take over another male's female. If such challenges occur during breeding, the female often cannot feed during her few minute incubation breaks, the only time in the day that she spends off the nest. She is either trying to deter her new admirer or is chased away from food sources by other flock members when her mate is busy fending off the competitor. As a result, the female will often face the choice of starving or abandoning the nest. This is typically decided in favour of survival. If an attempted takeover happens after hatching of the goslings, mother and goslings spend so much time avoiding the commotion that the goslings die within days, most likely due to exhaustion caused by too little feeding and brooding time. While these examples are certainly rare and unpredictable exceptions, they illustrate how severe the consequences of being overly aggressive, being forced to act aggressively, or just being in the way of aggressive individuals can be.

Data from a range of goose species, including the recent results from our greylag goose flock, thus depict similar benefits of dominance and aggression as those reported for many other species of animals, namely increased fitness via enhanced survival and reproduction. They also provide indications for potentially severe costs and for a possible variation in the costs and benefits between the sexes. One of our forthcoming tasks will be to perform analyses that are comparable in detail to those recently conducted on the causes of dominance and aggression. Combining our long-term data sets on dominance and individual life histories with state-of-the-art methodologies will help us to identify trade-offs between costs and benefits of aggressive behaviour, between the sexes and/or between life history stages, and will ultimately improve our understanding of the evolutionary trajectories of dominance-related behaviours in social vertebrates.

SUMMARY

Dominance and aggression are omnipresent in goose societies. How aggressive individual geese are and who wins interactions thereby depends on a number of factors related to an individual's personality (Chapter 3) and life history, the time of year and the social environment. Furthermore, goose parents influence their offspring's aggressive behaviour through genetic and non-genetic components that act in a complex age- and sex-dependent manner. Achieving a high dominance rank promotes juvenile survival, early pair bond formation and successful reproduction, but overt aggression may also impair reproduction and survival. Depending on the costs and benefits, optimal choices for achieving or maintaining a high dominance rank thus vary considerably between life history stages.

8

The costs of sociality measured through heart rate modulation

For decades, the benefits and costs of group living were a central topic in the field of behavioural biology (Rubenstein 1978; Krause & Ruxton 2002). Besides the wide-ranging benefits such as predator avoidance (Hamilton 1971; Treherne & Foster 1980; Fels *et al.* 1995), increased foraging efficiency (Clark & Mangel 1986; Creel 2001), and a wider choice of partners (Emlen & Oring 1977), it is also well known that group living bears costs, for example competition during foraging (Ranta *et al.* 1993; Barton & Whiten 2002) and mating (Emlen & Oring 1977), or parasite transmission (Côté & Poulinb 1995; Hughes *et al.* 2002). Most relevant for our work, social interactions between individuals within a group are one of the main sources of stress (hereafter referred to as 'social stress'), affecting an individual's physiology, behaviour, fertility and immune system (von Holst 1998; DeVries *et al.* 2003; McEwen & Wingfield 2003). The two major physiological systems involved in the mediation of stress responses are the sympatho-adrenomedullary system (SAS), which triggers the release of adrenalin into the bloodstream and innervates virtually every organ in the body, including the heart (Sapolsky 1992), and the hypothalamic–pituitary–adrenocortical (HPA) axis, which stimulates the adrenal cortex to release glucocorticoids such as corticosterone and cortisol (Siegel 1980). Whereas the activation of the SAS system works almost instantaneously, the synthesis of glucocorticoids and their release into the bloodstream takes seconds to minutes (Sapolsky 1992).

In the case of greylag geese, much is known about how social stress affects them in the long term and what criteria allow the ability

The Social Life of Greylag Geese: Patterns, Mechanisms and Evolutionary Function in an Avian Model System, ed. I. B. R. Scheiber *et al.* Published by Cambridge University Press. © Cambridge University Press 2013.

to cope better with social stress. For example, it is known that the stress load of parental individuals and singletons varies over the year (Kotrschal *et al.* 1998a), that reproductive success is greater in pairs with high testosterone compatibility (Hirschenhauser *et al.* 1999b; Weiß *et al.* 2010b; Chapter 4), and that the presence of a social ally significantly affects the hormonal stress response (Frigerio *et al.* 2003; Scheiber *et al.* 2005a, 2009a; Chapter 9). In order to fully understand physiological stress responses to social factors, it is important to study the modulation of both the HPA and SAS axes (Kralj-Fišer *et al.* 2010a). Examining heart rate (HR) enables us to measure the impact of single short-term events, such as brief social interactions which last for only a few seconds. Generally, it is known that short-term activation of the stress response is adaptive in coping with challenges, although it may be energetically costly (Selye 1950; Bartolomucci *et al.* 2005). In the 'acute stress response' or 'fight–flight response', energy is mobilised following exposure to a stressor, which allows an individual to adequately deal with it (Sapolsky 1992). In contrast, chronic and repeated activation of the stress axes, also termed 'allostatic load' or 'allostatic overload' (McEwen & Wingfield 2003), may have pathological consequences for an organism (Bartolomucci *et al.* 2005). Therefore, in terms of an individual's fitness and well-being, it is of major importance to respond to stressors appropriately (Nephew *et al.* 2003). Hence, one would expect animals like greylag geese, living in a complex social environment, to thoroughly evaluate single stressful events and flexibly modulate their physiological stress response accordingly. In order to measure this flexible modulation, HR presents an ideal tool, but measuring HR presents a methodological challenge. While it is comparably easy to measure stress hormone metabolites non-invasively from droppings, this is impossible for HR. We therefore implanted 25 greylag geese with HR transmitters allowing beat-to-beat HR recording, resulting in a huge and unique data set.

8.1 IMPLANTATION AND TRANSMITTER TECHNOLOGY

The chosen transmitters were fully internal, without an external antenna or repeater, in order to amplify the signal, and had a battery lifetime of approximately 18 months. The technical solution was developed by the Biotelemetry Group at the Research Institute of Wildlife Ecology of the University of Veterinary Medicine Vienna (Prinzinger *et al.* 2002). The transmitters weighed about 60 g, which is approximately 2% of the body weight of a greylag goose. This weight to body mass ratio is generally recommended in the literature, as such lightweight transmitters are not expected to handicap the animals in any way in their daily life (Dzus & Clark 1996). According to the objectives of the study, instantaneous HR on a beat-to-beat basis was transmitted, and 2-minute means of HR data were saved on an internal data logger over the entire period of 18 months. However, not all the transmitters worked for the full 18 months, due to technical difficulties (Table 8.1). Heartbeats were counted by a microcontroller and modulated into a pulse-interval signal of carrier frequency pulses. The transmitted pulses (duration 10 ms, radio frequency (RF) power about 4 mW) were short compared with the interval between the pulses and only every second heartbeat was transmitted, thereby doubling the life of the batteries. This is important, as the battery is the largest and heaviest part of the equipment in such a transmitter (the smaller the battery, the lighter the transmitter). The resulting decreased resolution of ±2 beats/minute was acceptable for the purposes of the present study. The equipment was adjusted to record and transmit data in the range of 50–500 beats/minute.

Implantation of the transmitters was performed by a team from the Veterinary University of Vienna led by Dr. Wolfgang Zenker, in five sessions from May to September 2005. After surgery, geese were housed individually in transport boxes overnight to allow them to recover from anaesthesia. After veterinary inspection, all 25 individuals were released into the flock the morning after implantation. All geese returned to, and were accepted by, their social partners immediately. One female disappeared the night after she was released and may have been taken by a fox, the main natural predator of geese at the study site. By 2–7 days after the surgery, the geese had fully recovered from surgery. They could not be told apart from non-implanted geese in appearance, behaviour or ability to fly, and had normal body temperature and HR. The implantation was conducted under an animal experiment licence issued by the Austrian Ministry of Science.

8.2 DATA COLLECTION

Data collection started, at the earliest, three weeks after surgery in order to exclude any possible effects of the operation on behaviour or HR. Data were taken on a daily basis if weather conditions permitted the use of the technical equipment (laptop, receiver). An observer simultaneously recorded HR and behaviour of a focal goose. All occurring behaviours of the focal individual were recorded and

Table 8.1 *Sex (M = male, F = female), year of hatching (hatch year), rearing condition (Gr = goose-raised, Hr = hand-raised), implantation date in 2005, body mass at implantation (g) and transmitter lifetime (months).*

	Sex	Hatch year	Rearing condition	Implantation date	Body mass	Lifetime
Armando	F	2000	Hr	15.08.	2,580	12
Balu	F	1993	Hr	29.07.	2,988	12
Blossom	M	2004	Gr	14.09.	3,606	15
Boston	M	2004	Gr	14.09.	3,778	17
Celine	F	2000	Hr	14.09.	3,732	11
Corrie	M	1999	Hr	12.08.	3,200	17
Edes	M	2000	Hr	29.07.	3,500	13
Gantenbein	F	2001	Hr	18.05.	2,960	11
Halas	M	2000	Hr	12.08.	4,400	11
Jacky	F	1999	Hr	11.08.	2,800	11
Jana	F	1998	Hr	14.09.	3,060	11
Jesaja	F	2003	Gr	11.08.	3,300	11
Juniper	M	2004	Gr	29.07.	2,842	13
Keiko	M	1996	Hr	14.09.	4,042	11
Lanzelot	F	2003	Gr	18.05.	3,088	3
Little	M	2004	Gr	12.08.	2,950	12
Loki	M	2001	Hr	18.05.	3,624	5
Lorelei	F	2003	Gr	11.08.	3,060	Disappeared August 05
Löwenherz	F	2001	Hr	18.05.	3,378	5
Sinus	M	1992	Gr	29.07.	3,264	1
Smoky	M	2001	Hr	18.05.	3,552	8
Spiro	M	2000	Hr	29.07.	3,566	Disappeared April 06
Terri	M	2000	Hr	29.07.	3,950	14
Tian	M	1999	Hr	12.08.	3,472	Disappeared March 06
Tristan	M	1999	Gr	11.08.	3,780	12

classified into categories. In addition, other social (aggressive encounters, socio-positive behaviours toward the focal individual) and non-social (cars passing by, thunderstorms) events that occurred in the vicinity of the focal individual were noted.

At any given time, the observer could only record HR for one focal individual goose. For this purpose, the human observer was standing at a maximum distance of 10 m from the focal individual,

Figure 8.1   Illustration of data recording in the field. A human observer (C.A.F.W.) carried a laptop computer on a special support frame. Simultaneously with heart rate, the behaviour of the focal individual was recorded using the Observer 4.1® software by Noldus, 2002. Photographs © A. Braun (*left*), C. A. F. Wascher (*right, top*)
A full-colour version is included in the colour plate section.

with a laptop computer on a support frame around his/her shoulders. The laptop computer was connected to the receiver (Fig. 8.1). Over the course of the study, we recorded HR and behaviour during all hours of daylight, covering different contexts such as feeding and resting. In addition to the daily location and detailed behavioural information on focal individuals, complete life history data for each individual goose were available (see Chapter 1 for details), making this data set quite unique. Some general information about the 25 individuals implanted with radiotransmitters is given in Table 8.1. These additional factors could then be included in the statistical analysis to investigate how general characteristics of the focal individual, for example sex, age and weight at implantation, influence HR.

We conducted generalised linear mixed models (GLMMs), applying the restricted maximum likelihood (REML) procedure for repeated sampling with an unbalanced design. In all models we included first-order interaction terms in the model. We sequentially deleted fixed terms in order of decreasing significance. Only terms with $p < 1$ remained in the final model. Mean HR during different behaviours, locomotions and body postures, and distances to the partner and the group, was used for analysis.

8.3 HEART RATE IN GEESE: THE OVERALL PICTURE

Looking at the entire HR data set, HR in greylag geese was significantly influenced by behaviour, locomotion and body posture, distance to the pair partner, rank, pair bond status and weight at implantation (Table 8.2). The fact that type of behaviour affects HR came as no surprise (compare Wascher *et al.* 2008a). We found overall mean HR to be highest during agonistic encounters and vocalisations, two behavioural categories in a clear social context, compared with other behaviours in a standard flock situation, such as resting (Fig. 8.2). Locomotion and body posture also significantly influenced HR, although differences in HR, for example between standing versus walking, or passively floating versus active swimming, were rather small (Fig. 8.3). A substantial HR change occurred during take-off and during the first seconds of flight, where HR was highest, i.e. above 500 beats per minute (bpm; Fig. 8.4). Anyhow, differences in HR according to behaviour and locomotion basically showed that HR was significantly modulated by physical activity, which can be explained by increased energy expenditure (Tatoyan & Cherkovich 1972; Dressen *et al.* 1990; Arnold *et al.* 2004). Rearing condition, age, sex and date of data collection had no significant effect on HR (Table 8.2). This shows that HR is modulated quite flexibly according to the social environment of each individual and that more general life history traits, such as sex or age, play a minor role.

Over the course of the entire study, differences between the sexes were not significant. However, we found seasonal variations between males and females. During the mating season, a period that is characterised by intense aggression between males (see Chapter 7), males had higher HRs than females. During this time, males have to invest much more into the social domain, while females prepare for the energetically demanding challenges of egg production, laying and incubation. Differences between the sexes in HR are not only seasonal but also context-dependent; for example, females are less reactive to agonistic encounters than males, and only in females is HR modulated depending on the distance from the partner (Wascher *et al:* 2012). In a previous study, the rearing condition was shown to influence the slow stress axis. Hand-raised geese excreted lower levels of immuno-reactive corticosterone metabolites (CORT) in response to an experimental stressor than did goose-raised geese (Hemetsberger *et al.* 2010; Chapter 3). Here, we did not find a difference in HR between hand-raised and goose-raised geese. Rather, our results show that

Table 8.2 *Factors affecting mean heart rate (HR) over different behaviours, locomotions and body postures and distance to the partner in greylag geese. Results of the generalised linear mixed model are shown. Results in bold remained in the final model.*

	Wald statistic	Effect size	p-value
Date	0.05	0	0.818
Behaviour	**27.92**	**0.001**	**<0.001**
Locomotion and body posture	**201.89**	**0.021**	**<0.001**
Distance to partner	**11.48**	**0.004**	**<0.001**
Sex	0.26	0	0.610
Age	1.50	0	0.221
Pair bond status	**5.28**	**0.026**	**0.022**
Rearing condition	0.24	0	0.624
Weight at implantation	**38.23**	**0**	**<0.001**
Rank	**17.70**	**0.006**	**<0.001**

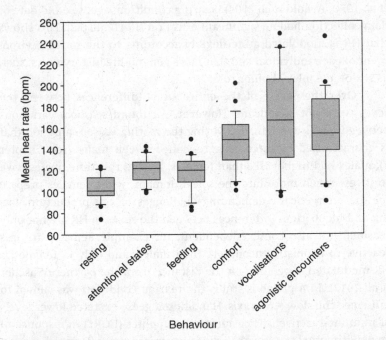

Figure 8.2   Mean heart rate in beats per minute (bpm) of 24 greylag geese in different behavioural categories. *Boxplots* show medians and quartiles, *whiskers* 10th and 90th percentiles, and *circles* indicate outliers.

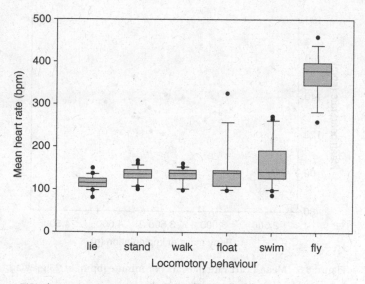

Figure 8.3   Mean heart rate in beats per minute (bpm) of 24 focal individuals during different locomotory behaviours and body postures. *Boxplots* show medians and quartiles, *whiskers* 10th and 90th percentiles, and *circles* indicate outliers.

Figure 8.4   Take-off of a greylag goose. © C. A. F. Wascher.

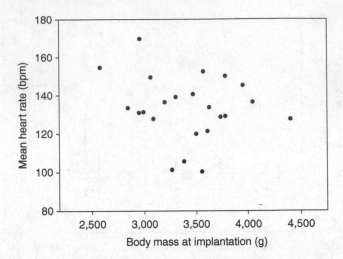

Figure 8.5   Mean heart rate in beats per minute (bpm) of 23 greylag geese and body mass on implantation day (g).

hand-raised and goose-raised individuals do not differ in their overall energy expenditure and respond in a similar way to challenges from the social environment, once more supporting our claim that socially involved hand-raising produces animals with species-typical behaviour (Hemetsberger *et al.* 2010; Chapter 1). In addition, body mass on the day of implantation influenced HR. Heavier individuals had a higher HR (Fig. 8.5). The relationship between body mass and HR in birds has also been described elsewhere (Grubb 1983). As well as body mass, rank (as estimated by the percentage of winning interactions) also influenced HR, with higher-ranking individuals having a higher HR than lower-ranking ones. The effects of dominance and social rank on the physiological stress response have previously been reported as concerning both the HPA (Creel 2001) and the SAS system (Candland *et al.* 1969, 1970). In our study, body mass and rank were correlated with each other (Spearman rank correlation: $r_s = 0.473$, $N = 23$, $p = 0.02$) and therefore these factors are not independent.

8.4 HEART RATE IN THE CONTEXT OF SOCIALITY

We found differences in HR depending on the distance to the partner, rank and pair bond status (Fig. 8.6) which could not be explained by differences in physical activity and associated increased energy expenditure. This could indicate HR modulation due to a lack of social

Figure 8.6   A greylag goose pair characterised by increased spatial proximity compared with non-associated flock members. © C. A. F. Wascher.

support. Social support is described as one of the key mechanisms for optimising the investment into the social domain and is defined as the stress-reducing effect of the presence of a familiar social partner (Sachser *et al.* 1998; von Holst 1998). The effects of social support on behaviour and CORT have been well described in greylag geese (Frigerio *et al.* 2003; Weiß & Kotrschal 2004; Scheiber *et al.* 2005a, 2009a; Chapter 9). For the first time in non-human animals, our data indicate that social support might affect the HR component of stress as well. First, we found a context-dependent difference in HR between paired and unpaired greylag geese (Wascher *et al.* 2012). Especially during socially stressful agonistic encounters, HR in unpaired individuals was significantly higher than HR in paired individuals. This shows that individuals with a partner respond less to social stressors, leading to a possible optimisation of the energetic investment in the social domain. Apart from the presence or absence of a social ally, either a partner or family member, the distance from that social ally modulated HR as well. In females, HR increased when the social partner was further than 2 m away. Given the fact that greylag geese actively support one another in agonistic encounters (Scheiber *et al.* 2005a, 2009b; Chapter 9), a close proximity to one's social ally is desirable, so that he/she is able to actively interfere. Presumably the close proximity to one's partner may also have a calming effect which, in turn, results in lower HR, whereas a larger distance to a potential supporter evokes the opposite effect. Furthermore, the fact that the mere presence of a social ally has stress-reducing effects is corroborated by the HR results.

In addition to the effects of the social environment on HR, we were particularly interested in the effect of single agonistic events, and whether HR was modulated flexibly according to the relevance of such events for the focal individual. Furthermore, we investigated whether HR in greylag geese was modulated mainly by the metabolic demand of such events (Obrist 1981) or by an individual's psychological state (Blix *et al.* 1974). Investigating HR modulation in response to single agonistic encounters, we found that the characteristics of both the interaction and the opponent itself influenced HR. The former reflects HR modulation, which may be the result of increased physical activity. The more intense an encounter and the longer it lasts, the higher the HR. HR modulation in response to the opponent, on the other hand, cannot be explained by increased energy expenditure, but rather by motivational factors; a finding that could not have been predicted from the behaviour of the animals (e.g. intensity of the interaction) alone. Generally, HR increased to a greater extent when the risk of being defeated was higher, i.e. in interactions with a higher-ranking individual (Wascher *et al.* 2009). This may be seen as an individual stress-coping strategy; the increased HR in response to an agonistic encounter is part of the initial 'alarm phase' of the stress response (Axelrod & Reisine 1984; Stoddard *et al.* 1986). This physiological reaction prepares the body to cope with a stressful event. Such an activation of the stress response is connected with considerable energetic costs, and an inappropriate activation has negative effects on an individual's energy reserves and ultimately on its reproductive success (Bartolomucci *et al.* 2005).

Our results suggest that individuals modulate their HR very flexibly according to the relevance of single agonistic encounters, and this is determined mainly by the risk of being defeated. If this risk is high (i.e. when interacting with a higher-ranking individual) the stress response is up-regulated even in anticipation of such an event. When the risk of losing is low, any additional energy pre-investment towards winning the contest is unnecessary and HR increases remain moderate. This also illustrates that in this bird species, besides physical activity, psychological factors are a driving force of HR (Wascher *et al.* 2008a). We can only speculate about possible long-term consequences of HR modulation to social factors. It can be expected that adequate modulation of HR due to social factors remains without negative long-term

Figure 8.7   Female greylag goose on incubation break calling for her partner. © C. A. F. Wascher.

consequences for the individuals. In any case, frequent and long-term elevations in HR are energetically costly, with negative effects on an individual's fitness (von Holst 1998; Bartolomucci *et al.* 2005).

It is particularly fascinating that not only active involvement in agonistic encounters influences HR, but also the mere observation of relevant social interactions without any discernible changes in behaviour and/or locomotion (Wascher *et al.* 2008b). HR responses of bystanders differed depending upon the individuals involved: geese responded more strongly when watching interactions with a socially relevant individual, such as the pair partner or a family member, as compared with events in which non-associated individuals were involved (Wascher *et al.* 2008b). A similar pattern appeared when comparing HR responses to conspecifics' vocalisations: HR was significantly higher when the calls of a pair partner, isolated from the flock during an experiment, were played back, than when calls of non-associated geese were projected (Fig. 8.7; I. Nedelcu *et al.*, unpublished). These results not only show that greylag geese can obviously vocally discriminate between the pair partner and non-associated geese, but also that 'bystanding' results in a specific arousal, reflecting the motivational/emotional state of that individual.

This suggests that greylag geese evaluate social events according to their relevance and modulate their physiological stress response accordingly. When the pair partner or a family member is involved in an interaction, it is much more likely that the focal individual will interfere (Scheiber *et al.* 2009b) and provide active social support than when non-associated geese interact. Therefore, the higher HR responses of bystanders following socially relevant interactions can probably be seen as bodily preparation for any action appropriate or necessary in that particular context. HR has been suggested as a meas ureable parameter to assess affective arousal in humans (McCraty *et al.* 1995; Scherer 2005) as well as in non-human animals (Aureli *et al.* 1999; Bradley & Lang 2000; Cabanac & Cabanac 2000; Bradley *et al.* 2001; Preston & de Waal 2002; Desire *et al.* 2004). HR increases in response to relevant stimuli in the absence of physical activity have been anec- dotally reported from bighorn sheep (*Ovis canadensis*: MacArthur *et al.* 1982), herring gulls (*Larus argentatus*: Kanwisher *et al.* 1978) and guide dogs (*Canis lupus familiaris*: Fallani *et al.* 2007). Furthermore, affective responses, including physiological, behavioural and subjective com- ponents, are a result of the cognitive, not necessarily 'conscious', evaluation of an event in terms of its significance for the survival and well-being of the individual (Scherer 2005; Desire *et al.* 2006).

In sum, we found that the social setting has pronounced effects on an individual's physiology. This may have serious consequences with respect to energy expenditure, because HR provides a fair approx- imation of oxygen consumption (Ward *et al.* 2002). As a consequence, HR is modulated in response to relevant social events. We have shown that living in social groups is energetically costly, but the individual's investment is optimised by flexible modulation of the physiological stress response according to subsequent needs and requirements.

SUMMARY

Social contexts are among the most potent stressors, affecting individ- ual behaviour, physiology and the immune system. Although energeti- cally costly, short-term elevation of the stress response is adaptive and unavoidable in coping with challenges. In contrast, chronic activation of the stress axes may have pathological consequences. Therefore, in terms of an individual's fitness and well-being, it is of major impor- tance to respond to stressors adequately. However, socially induced stress does not affect all individuals of a population in the same way. Depending on their coping style, social status and embedding, individuals differ in their physiological responses to social stressors and other challenges. Modulation of glucocorticoids and heart rate (HR) are direct indicators of an individual's total stress load and the associated energetic costs. In this chapter, we present data from 24 greylag geese which were implanted with sensor-transmitter packages for 18 months in order to measure short-term HR responses to single social events. The general characteristics of an individual such as sex and age had no significant effect on HR, whereas body mass at the time of implantation correlated positively with HR. HR increased substantially during extreme physical activity, such as take-off, and in addition, social interactions proved to be the strongest HR modulators. During agonistic encounters, HR was increased according to the characteristics of the interaction and the opponent. These results showed that individuals modulate their HR very flexibly according to the relevance of single events. This adjustable modulation of the stress response may serve as a mechanism to keep the energetic costs of social life as low as possible, so that physiological investment takes place only when it matters. This helps individuals to optimise the costs associated with social life, as they evaluate every social and non-social event in their environment and modulate HR according to this evaluation.

9

'Tend and befriend': the importance of social allies in coping with social stress

Von Holst (1998) describes stressors as 'all social and non-social stimuli that are challenging or threatening to the survival, health, and reproductive success of animals and that are, therefore, an essential part of natural selection'. Social interactions are actually among the most potent stressors (DeVries *et al.* 2003; Chapter 8). On the other hand, social contact with allies as well as affiliative behaviours, such as reconciliation or consolation after conflict (Schino 2000), may buffer stress, increase success in agonistic encounters, ease access to resources, and may have a positive impact on health and longevity (Christenfeld *et al.* 1997; DeVries *et al.* 2003). Depending on the social situation and individual behavioural phenotype, social challenges may affect individuals differently across all members of a group (Sapolsky 1994; Koolhaas *et al.* 1999; DeVries 2002; DeVries *et al.* 2003; McEwen & Wingfield 2003; Goymann & Wingfield 2004). Therefore, it is crucial to understand the means employed by individuals to gain advantages over others in the competition for food and reproductive success. One possibility is through the support of social allies, which is a common phenomenon in primates (Berman 1980; Datta 1983a, 1983b; Horrocks & Hunte 1983; Bernstein & Ehardt 1985, 1986; Netto & van Hooff 1986; Pereira 1992) as well as non-primate mammals such as alpine marmots (*Marmota marmota*), guinea pigs (*Cavia porcellus*) and tree shrews (*Tupaia belangeri*) (von Holst 1986a; Hennessy & Ritchey 1987; Arnold & Dittami 1997; Kaiser *et al.* 2003). However, this complex of mechanisms, such as 'tend and befriend' responses of females (Taylor *et al.* 2000) has hardly been investigated in birds.

The Social Life of Greylag Geese: Patterns, Mechanisms and Evolutionary Function in an Avian Model System, ed. I. B. R. Scheiber *et al.* Published by Cambridge University Press. © Cambridge University Press 2013.

Social support has been defined as the beneficial effect of the presence of a familiar social partner (Sachser *et al.* 1998; von Holst 1998). Two kinds of benefits from social alliances, which are not mutually exclusive, have been defined: 'active social support', which describes the participation of a social ally in agonistic encounters (Horrocks & Hunte 1983; Pereira 1992), and 'passive social support' through the stress-reducing effects of the mere presence of a social ally (von Holst 1986a; Sapolsky 1992; Levine 1993a, 1993b; Sachser *et al.* 1998; Smith *et al.* 1998). In particular, the passive supportive effects of social allies have been little studied and are almost unknown in birds (Dugatkin 1997).

Active support by social allies in agonistic encounters has been described for many mammalian species (Berman 1980; Horrocks & Hunte 1983; Datta 1983a, 1983b; Bernstein & Ehardt 1985, 1986; Netto & van Hooff 1986; Hennessy & Ritchey 1987; Pereira 1992; Arnold & Dittami 1997; von Holst 1998; Smith *et al.* 2010), as well as for a few birds, mainly geese (Boyd 1953; Hanson 1953; Lazarus 1978; Black & Owen 1987; Ekman *et al.* 2000; Stahl *et al.* 2001). Active support may mean either physical involvement in agonistic actions, or any type of vocalisation, threatening and reconciliation displays during, and possibly after, aggressive encounters.

Studies linking support by social allies to physiological parameters, i.e. passive social support, manifested in the reduction of stress parameters, have only recently been conducted, although the stress-reducing effects of social allies were predicted to influence both the fast and slow stress axes. With two exceptions (Fokkema & Koolhaas 1985; von Holst 1998), no information is available on catecholamine baseline values as a measure of the sympatho-adrenomedullary 'fast' stress response (SAS) in animals in social situations. However, continuous long-term measurement of the heart rate provides a good indicator for this activation (von Holst 1986b; Eisermann 1992; Chapter 8). The social environment is known to modulate heart rate substantially. For example, crowding as a social stressor has been shown to increase heart rate in European starlings (*Sturnus vulgaris*: Nephew & Romero 2003). Agonistic interactions raised heart rate in herring gulls (*Larus argentatus*: Kanwisher *et al.* 1978), and the heart rate in one greater white-fronted goose (*Anser albifrons*) increased from 80 beats per minute to over 400 beats per minute during an aggressive encounter (Ely *et al.* 1999). Wascher *et al.* (2009) found similar responses in greylag geese (see also Chapter 8).

Social interactions and social status are also known to modulate the 'slow' stress response, i.e. the hypothalamic–pituitary–adrenocortical axis (HPA) in mammals and birds (Wingfield & Silverin 1986; Kotrschal et al. 1998a, 2000; Creel 2001; DeVries 2002; DeVries et al. 2003). The close proximity of social allies during and after a stressful event, however, may dampen this physiological response. This phenomenon is known as passive social support. In humans, passive social support is sometimes also referred to as 'emotional support' (Berkman et al. 1992; Hamre & Pianta 2005).

In this chapter, I give an account of our research on social support in greylag geese, which revealed many aspects that resemble social support in mammals, including primates. Their complex social system (Rutschke 1982; Schneider & Lamprecht 1990; Fox et al. 1995; Kotrschal et al. 2006, 2010) makes greylag geese a promising avian model to study social support, as the system features the three conditions considered necessary by von Holst (1998) for the development of social support systems in mammals:

1.   a complex social organisation (Weiß et al. 2008; Kotrschal et al. 2010; Chapter 11)
2.   long-term relationships with bonding partners (Chapter 4)
3.   female-centred clans (Frigerio et al. 2001a; Chapter 6).

Adopting an integrated approach, we studied the impact of social support on behaviour, physiology and its effects on the greylag goose social system. The main objective of our studies was to pinpoint the role of social support as a prime mechanism of individual stress management in freely moving, socially intact greylag geese. We were particularly interested in the effects of passive social support (Frigerio et al. 2003; Weiß & Kotrschal 2004; Scheiber et al. 2005a, 2009a). We also wanted to determine the distribution of active and passive support provided and received by different social allies, i.e. males, females and goslings/juveniles (Scheiber et al. 2005a, 2009a, 2009b; Swoboda 2006), and were interested in the role of family size on social support, and thus in the possible benefits of acceptance of unrelated young ('adoption': Choudhury et al. 1993; Kalmbach et al. 2005; Scheiber et al. 2005a).

It is noteworthy that although living in stable social units can reduce stress, social engagement is also costly and can cause stress. Evolutionary theory predicts that there is a trade-off between the benefits and costs of living socially, as well as a conflict between parents and offspring (Alexander 1974; Trivers 1974) as to when to terminate family bonds. From an evolutionary perspective, one should break the social unit/family bonds if the individual costs of maintaining a social unit or staying together as a family are higher than if the bond is broken. The stable family bonds in greylags indicate that the benefits will outweigh the costs over evolutionary time, and therefore have been selected for the relevant physiological, behavioural and psychological mechanisms and dispositions.

9.1 MEASURING SOCIAL SUPPORT IN THE KLF GEESE

Over the years, we have performed our studies on active and passive social support with both hand-raised and goose-raised greylag goose families. To determine the effects of active social support we collected behavioural observations during several different seasonal phases:

- re-establishment of the flock after moult in the autumn, when the dominance hierarchy is reinstated
- the stable winter flock, when the dominance hierarchy is well established
- disintegration of the flock into pairs and loosening of the parent–offspring bonds in early spring, due to the pairs preparing for breeding.

For the hand-raised geese, data were also collected during the second autumn, winter and spring, when the hand-raised subadult geese re-enter the flock (Frigerio *et al.* 2003; Weiß & Kotrschal 2004). This is comparable to goose families rejoining their parents to form a secondary family (see Chapter 6). For the goose-raised geese, we collected similar data for ten primary families (Scheiber *et al.* 2005a) as well as ten secondary families (Scheiber *et al.* 2009a).

In general, we observed agonistic interactions of focal individuals either during the morning feedings (08:00 h) or the afternoon feedings (14:00–17:00 h, depending on the season). During feedings, social stress can easily be manipulated via the distribution of the food: spreading it over the whole feeding area (approx. 150 m^2) results in a relatively relaxed control situation, while spreading the same amount over approximately a quarter of that area (approx. 40 m^2) results in competitive 'social density stress' feeding ('SDS': Kotrschal *et al.* 1993; Scheiber *et al.* 2005b). Observations ended when the focal individuals left the feeding area, or after one hour – whichever occurred first. Behavioural protocols consisted of continuous focal sampling of agonistic interactions with other flock members.

An agonistic interaction was defined as an encounter between two geese in which one of them evoked withdrawal behaviour from the opponent. The withdrawing goose was considered the loser in that interaction; the goose which evoked the withdrawal was considered the winner. During the observation we recorded which focal individual was involved in an agonistic interaction, the social category (family, paired, single) and the sex of the opponent, as well as the distances between the focal individual and its family members. In the case of goose-raised geese, we also noted whether the interaction was won or lost with or without active social support. We defined interactions won or lost with active support if at least one family member assisted the focal individual during an interaction (Scheiber et al. 2005a). This assistance was often a supportive shouting display by the rest of the family, with each individual stretching its neck forward towards the opponent (Lorenz 1988). We also noted whether the original agonistic encounter was followed by another attack by the family against the same opponent ('repeated agonistic attacks': Scheiber et al. 2009b).

The behavioural protocols of hand-raised focal individuals were collected in the control situation only, but in the presence of different humans. The humans had different social relationships with the focal animals and were either the human foster parent (treatment A), or a familiar human (the human foster parent of another sibling group; treatment B), or a known but non-familiar human (treatment C) who was generally a person whom the focal goose might have seen before, but to whom it was not individually habituated. In a fourth treatment (D), no human was present. In addition to recording agonistic interactions, we also recorded instantaneous focal sampling of feeding at 1-minute sample intervals.

In addition to behavioural observations, we collected droppings for quantitative analysis of immuno-reactive corticosterone metabolites (CORT) of all individuals within the focal family for a defined period of time. Sampling started sufficiently late in the day to avoid the endogenous CORT early morning peak (Schütz et al. 1997). However, the sampling protocol varied slightly between studies. Frigerio et al. (2003) collected one sample per individual within 90 minutes after the morning behavioural data collection, when CORT excretion usually peaks in geese following a challenge, e.g. an injection of adrenocorticotrophic hormone (ACTH) or the presence of unfamiliar individuals in the group of geese (Krawany 1996; Hirschenhauser et al. 2000). Scheiber et al. (2005a; 2009a) collected a sample of at least three or four droppings within 3 hours after feeding had started. We had shown previously that by increasing the number of droppings collected, the pattern of CORT in geese, where concentrations vary widely from one dropping to the next, could be measured more precisely and an individual's acute stress response could be inferred reliably even when collecting excreta rather than blood (Scheiber *et al.* 2005b). After collection, samples were frozen and droppings were subsequently assayed with enzyme immunoassay ('EIA': Möstl *et al.* 1987; Kotrschal *et al.* 1998a, 2000). In earlier studies, a group-specific antibody, which recognised 11β,21OH,20-oxo-corticosterone metabolites, was used (Frigerio *et al.* 2003; Kralj-Fišer *et al.* 2007). Later on, a new group-specific antibody was developed (Möstl *et al.* 2005), recognising 5β,3α,11β-diol glucocorticoid metabolites, which proved considerably more sensitive, resulting in higher peak values (Frigerio *et al.* 2004b). This 'new' assay was used in all later studies (Frigerio *et al.* 2004b; Scheiber *et al.* 2005a, 2005b, 2009a).

9.2 ACTIVE SOCIAL SUPPORT IN GREYLAG GEESE

Active social support was relatively common in primary families with their fledged offspring (Fig. 9.1). However, the percentages of social support received differed significantly between family members (one-way ANOVA: $F = 5.580$, $N_{fathers} = 10$, $N_{mothers} = 10$, $N_{sons} = 13$, $N_{daughters} = 20$, df = 3, $p = 0.002$). In particular, sons received more active social support than either their fathers (Holm–Sidak $p < 0.01$) or mothers ($p = 0.01$), but no such relationship was evident in daughters. Also, social support decreased over time: while it was relatively common in summer and winter (38.6% and 44.6% of agonistic interactions, respectively), active social support practically disappeared towards the start of the mating season, when families are about to break apart (6.5% of agonistic interactions; Scheiber *et al.* 2005a). All members of the family provided active social support (vocal and/or physical) to each other, but while both parents won two-thirds of their interactions without active support, the juveniles won more interactions with than without active support (Scheiber *et al.* 2005a). Whereas family size did not influence the occurrence of social support, it influenced the per capita number of agonistic interactions: the larger the family, the fewer the agonistic encounters they were involved in (Scheiber *et al.* 2005a). This was most pronounced during social density stress feedings, but was also apparent in control situations (Scheiber *et al.* 2005a).

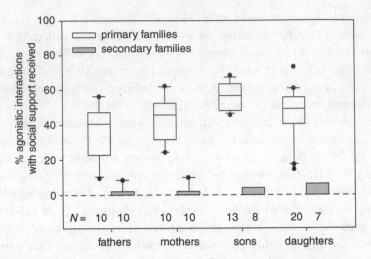

Figure 9.1:   Percentage of agonistic interactions with social support received by fathers, mothers, sons and daughters in primary (*white*) and secondary (*grey*) goose families. *Boxplots* show medians and quartiles, *whiskers* 10th and 90th percentiles, and *circles* indicate outliers.

During this time, juveniles also benefit from another form of active social support, the occurrence of repeated aggression by family members against a former target. After a juvenile was defeated, family members chased the juvenile's opponent significantly more often than expected by chance (Chi-square test: χ^2 = 2,042.74, df = 8, p < 0.001), where 28% of the redirected attacks were performed by the father, 40% by the mother, 24% by the siblings, and 8% by the primarily attacked juvenile itself (Scheiber *et al.* 2009b). In addition, juveniles chased other flock members significantly more often than expected after a sibling had been displaced (Weiß 2000). These repeated attacks were relatively common; in 10.4% of agonistic interactions performed by ten different families, immediate attacks by family members against the original target followed. We observed up to five attacks of family members against the same target, with one being the most common (86%; Scheiber *et al.* 2009b). The chance of winning such a follow-up attack was higher than winning the initial attack (93% versus 71%).

Supporting family members either directly in fights (Scheiber *et al.* 2005a) or through repeatedly attacking a formerly successful opponent (Scheiber *et al.* 2009b) not only signals strong agonistic potential and high rank to other members of the flock. It also reverses previous losses, reinforces a losing experience on the side of the opponent and a winning experience on the side of the family. Indeed, those juveniles that were most involved in serial attacks while still in the family unit had a higher dominance rank after the family break-up (Scheiber *et al.* 2009b).

Active social support between goslings and parents becomes relevant only towards fledging of the goslings, and in particular when the flock starts to re-aggregate after moulting, as it appears rather late during development (Swoboda 2006). This suggests that social support is a key requisite of the extended family bonds, but of little or no relevance while goslings are fully dependent on their parents at a time when they are no match for other flock members in body size (see also 'gosling helper hypothesis': Black *et al.* 2007). Once goslings hatch, males are involved in most of the agonistic encounters of the family, and as parental males are top-ranking in the flock (Weiß *et al.* 2011), they win agonistic encounters without support (Swoboda 2006). Agonistic interactions between goslings and other flock members and active help by family members are almost entirely absent until the goslings are 6–7 weeks old. From then on, goslings gradually increase their involvement in aggressive interactions until the flock reunites after moulting in the autumn (Swoboda 2006). Goslings win these encounters with the active help of their fathers even against larger and older individuals, but frequently lose when their parents are not nearby (Weiß & Kotrschal 2004; Swoboda 2006).

The previously described studies demonstrated the importance and benefits of social support in extended family units until the families break up in spring (primary families). If parents fail to reproduce in their next breeding attempt the following year, the young of previous years might rejoin their parents as a secondary or even tertiary family (Lorenz 1988; Chapter 6). While it may be assumed that primary families after fledging and secondary families form extended bonds for the same reasons, the mechanisms behind, and benefits of, reuniting families had not previously been studied in detail. We also investigated social support in these contexts (Scheiber *et al.* 2009a). Active social support was virtually absent in secondary families (primary families: 38% of 4,474 interactions; secondary families: 1.5% of 965 interactions: Scheiber *et al.* 2009a; see also Fig. 9.1). The latter, therefore, more probably form because of motivational components (Lamprecht 1986a) as well as the benefits gained from passive social support. Although secondary families rank lower in the hierarchy than do primary families, they may nevertheless outrank pairs that have not rejoined with their offspring (Chapter 6). We suggest that the reason why subadult males, in particular, join their parents may be attributed to these motivational factors (Lamprecht 1986a), as these males, and to some extent their sisters, will win agonistic encounters relatively often, thereby gaining higher rank than when outside the secondary family unit (Weiß et al. 2008; Scheiber et al. 2009a).

9.3 PASSIVE SOCIAL SUPPORT IN GREYLAG GEESE

9.3.1 Behavioural effects

Active and passive components are often difficult to tell apart in intact goose families because goose parents provide active support when needed. However, hand-raised juveniles and their human foster parents are suitable models for disentangling the active and passive motivational components of social support: human foster parents defend their goslings against other geese just like goose parents do, but they can be instructed not to do so at certain times, so that juveniles are supported only passively (i.e. due to the mere presence of the human foster parent). At the KLF, hand-raised sibling groups are generally accompanied by their human foster parents until well after fledging, but not over winter as goose parents do. While juveniles were found to be similar in dominance rank to goose-raised juveniles in the presence of a human foster parent, those without their human ally present ranked as low as single individuals in the flock (Weiß 2000), although their sibling groups were still complete. This condition is comparable to an early family break-up, as is sometimes observed in wild barnacle goose (*Branta leucopsis*) families (Black et al. 2007). Hand-raised juveniles enhanced their agonistic success (Fig. 9.2) and fed significantly longer in the presence of their human foster parent or a well-familiarised human, compared with situations when a non-familiar human or no human was present during the feeding of the flock (Fig. 9.3; Weiß & Kotrschal 2004). As neither human present interfered actively in the juveniles' agonistic interactions, their enhanced success in the presence of human allies may thus be attributed to the behavioural effects of passive support alone. Similarly, at times when the family unit is still intact, but the parent is temporarily 'unavailable', juveniles experience an immediate drop in dominance rank, which is also seen in the offspring of goose families that are temporarily displaced. Furthermore, this set of experiments indicates that greylag geese not only distinguish between individual humans, but also that they distinguish between those who previously provided active support for them and those who did not.

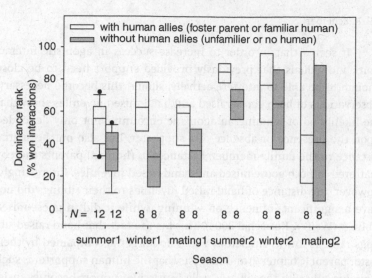

Figure 9.2 Dominance rank of hand-raised geese from fledging to adulthood in the presence (*white*) or absence (*grey*) of human allies. Seasons marked with '*1*' refer to the first year of data collection (juvenile geese), seasons marked with '*2*' to the second year (subadult geese). *Boxplots* show medians and quartiles, *whiskers* 10th and 90th percentiles, and *circles* 5th and 95th percentiles.

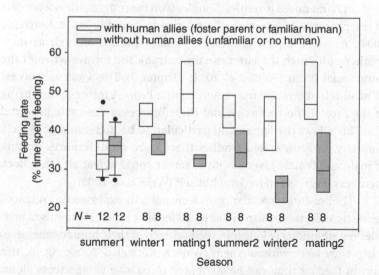

Figure 9.3 Feeding rates of hand-raised geese from fledging to adulthood in the presence (*white*) or absence (*grey*) of human allies. Seasons marked with '*1*' refer to the first year of data collection (juvenile geese), seasons marked with '*2*' to the second year (subadult geese). *Boxplots* show medians and quartiles, *whiskers* 10th and 90th percentiles, and *circles* 5th and 95th percentiles.

It seems that in order to increase success in agonistic interactions, individuals that previously provided support need to be close enough to be able to interfere actively, should this become necessary. Observations of both goose-raised and hand-raised juveniles show that the likelihood of winning an agonistic encounter not only depended upon the presence or absence of a supporter, but also on the spatial distance to the family member. Distances to (human) parents thereby mattered in both goose-raised and hand-raised juveniles. Interestingly, however, the distance of hand-raised juveniles to their siblings did not have a significant influence on winning, while it did in goose-raised siblings (Weiß & Kotrschal 2004). This fits the fact that hand-raised siblings rank as low as singletons when no longer accompanied by their foster parent (Chapter 7) and suggests that the human supporter is such a mighty ally that the siblings' role in passive support becomes insignificant. Similarly, the agonistic success of goose-raised juveniles was higher when in close proximity to higher- rather than lower-ranking family members (Weiß 2000). This indicates that the higher-ranked the family members were, the more effective they were as supporters. Also, the young of larger families had fewer agonistic interactions per capita than juveniles from small families (Scheiber et al. 2005a).

In hand-raised juveniles, females won more agonistic encounters in the presence of a human ally relative to males (Weiß & Kotrschal 2004). As a consequence, juvenile females ranked as high as their brothers, although the latter usually outrank the former when of the same social status (Weiß et al. 2011; Chapter 7). The agonistic success of hand-raised females may result from a human foster parent being in the fore of the hierarchy, and therefore being the most powerful ally. This allows the females, in particular, to be high-ranking as well. Similarly, in goose-raised families there were no differences in rank of male and female juveniles if the father could defeat all other flock members; i.e. was top-ranking himself (Weiß et al. 2008).

The benefits of social support were not only expressed in enhanced agonistic success, but also in increased feeding times. Hand-raised juveniles fed significantly longer in the presence of their human allies than when they were without them (Weiß & Kotrschal 2004). Again, this gain in feeding time can be attributed to passive components alone, since the humans did not monopolise food sources or interfere in any other way. These results are in concordance with observations in barnacle goose juveniles, where those individuals that were still embedded in a family fed uninterrupted for longer periods of time than did those who had already split from their families (Black et al. 2007).

Data from hand-raised juveniles suggest that the passive components of social support may also be important later on. After a decline in the percentage of interactions won during their first spring, hand-raised juveniles managed to attain agonistic success rates comparable to those after fledging during their second summer when accompanied by their human allies. Unlike the first year, this effect declined towards winter despite the continuing presence of their human ally, and was no longer detectable in the next mating season, when juveniles reached sexual maturity. Notably, however, the positive effects of the human allies on feeding rates were maintained through to adulthood (Weiß & Kotrschal 2004).

9.3.2 Physiological effects

In addition to the behavioural effects of passive social support, we investigated the influence of the presence of social allies on the physiological stress response. In goslings, the concentrations of glucocorticoid metabolites (CORT) in droppings are generally highest shortly (i.e. up to 20 days) after hatching, and then continue to decrease thereafter (Frigerio et al. 2001b; Swoboda 2006). The lowest values are found shortly before (Swoboda 2006) and after fledging (Scheiber et al. 2005a). To determine the effects of social allies on CORT excretion, we exposed primary and secondary goose families to two conditions: (i) a control feeding and (ii) a social density stress feeding (SDS) one day thereafter. For a 3-hour period after each of these two feeding conditions, we collected the droppings from one focal family for the determination of CORT. After CORT extraction from these samples, we calculated a specific value of change in CORT concentration per goose. This value, termed Δ CORT, is calculated as the mean CORT concentration collected during control feedings deducted from the mean CORT concentration collected under a social density stress situation (for a detailed description of the method, see Scheiber et al. 2005a, 2009a). Δ CORT values can either be positive, indicating a higher CORT excretion during social density stress, equal to zero, or be negative, indicating a higher CORT excretion during control feedings (Scheiber et al. 2009a). The occurrence of these negative values may seem counterintuitive at first; however, these are the ones to indicate a social support effect – i.e. the reduction of the stress response relative to a standardised control.

We found elevated CORT values in both primary and secondary families during the control feedings relative to social density stress feedings, i.e. evidence of passive social support (Scheiber *et al.* 2005a, 2009a). Why is it that families seem to perceive a high-density 'stress' feeding to be less challenging than a low-density control feeding situation? While the increased density of food and thus, closer proximity of the flock, led to elevated CORT values in geese without offspring (S. Kralj-Fišer & I. B. R. Scheiber, unpublished), those of families with more than two offspring showed a decrease in their stress response. This is likely to be a consequence of the closeness of the social allies during the stress feedings, where a reduced effort in monitoring the family members' positions is possible. In the control feedings, on the other hand, small sibling subgroups roam the feeding area alone, and parents show constant vigilance in tracking their positions. For larger families, this seems to be a more potent stressor than the social density stress feedings.

When examining our findings more carefully, the reduced CORT response was more prominent in adult females than males (Scheiber *et al.* 2005a) but was also detectable in juvenile males and females (Fig. 9.4, white boxes). The effect in adult females, however, was only evident in larger families; females with at least three young seemed to benefit, whereas females from small families (1–2 young) did not show the decrease in CORT (Scheiber *et al.* 2005a). Furthermore, the finding is more pronounced in winter (Scheiber *et al.* 2005a), a time when females must start boosting their body reserves to be able to build eggs the following spring, as these are produced from stored lipids (Prop *et al.* 1984). Maternal females seem to gain benefits from the reduction in CORT during their next breeding attempt, because in geese successful reproduction increases the prospects for successful future reproduction (Cooke *et al.* 1995; Hemetsberger 2002).

The physiological benefits of passive social support were also evident in secondary families. Throughout the second year, adult males, females, subadult males and females had lower CORT values overall when they rejoined a family unit (Fig. 9.4, light grey boxes) than those who did not rejoin (Fig. 9.4, dark grey boxes). A significant reduction, however, was again found in both adult and subadult females only (Scheiber *et al.* 2009a), whereas the reduction in CORT during social density stress was detectable neither in adult nor subadult males. Both adult and subadult females seem to ensure the comparable physiological advantages found in primary families when forming a secondary family unit (Scheiber *et al.* 2009a).

CORT data from hand-raised juveniles showed a somewhat different picture than those from goose-raised individuals. Against

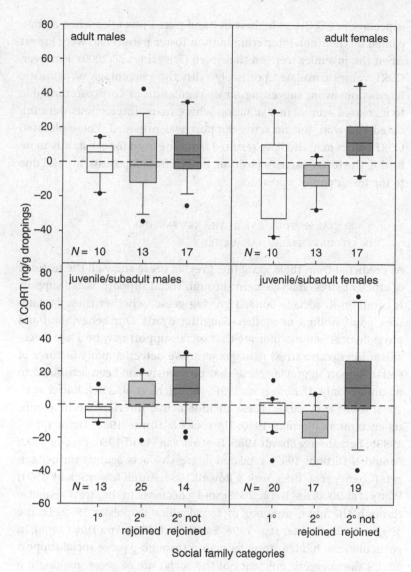

Figure 9.4 Δ immuno-reactive corticosterone metabolites (CORT) of male and female parents and offspring in primary families (1°, *white*), secondary families (2°, *light grey*) or when not rejoined as a secondary family (2°, *dark grey*). Values *above the dashed line*, which indicates equal CORT excretion during social density stress and control (zero line), indicate higher CORT values during social density stress feedings, values *below the dashed line* indicate higher CORT values during control feedings. *Boxplots* show medians and quartiles, *whiskers* 10th and 90th percentiles, and *circles* indicate outliers.

expectation, CORT was highest in hand-raised juveniles when accompanied by their non-interfering human foster parent and were lowest when the juveniles were on their own (Frigerio *et al.* 2003). However, CORT values correlated positively with the percentage of agonistic interactions won, suggesting an up-regulation of corticosterone due to increased arousal in a situation where many interactions were initiated and won, but no active support was provided. These elevated CORT values may also have resulted from the need for constantly monitoring the foster parent's position, and/or from a positive arousal due to the foster parent's presence.

9.4 SOCIAL SUPPORT, ONE OF THE KEY FACTORS IN EFFICIENT STRESS MANAGEMENT

As predicted from their social life, greylag geese show all the features of efficient stress management through social support. Social support is common in socially bonded greylag geese, whether these be families, adult siblings or mother–daughter dyads. Our behavioural and physiological findings indicate that social support may be a key mechanism in effective stress management. We detected many features of social support in greylag geese that previously had been attributed to mammals only (Uchino *et al.* 1996; Heinrichs *et al.* 2003; Kaiser *et al.* 2003; Brent *et al.* 2011). These include active interference in agonistic encounters (Berman 1980; Horrocks & Hunte 1983; Datta 1983a, 1983b; Bernstein & Ehardt 1985; Netto & van Hooff 1986; Pereira 1992; Arnold & Dittami 1997), repeated aggressive acts against former targets (Cheney *et al.* 1986; York & Rowell 1988; Aureli & van Schaik 1991; Watts *et al.* 2000; Engh *et al.* 2005), and a decrease in the stress response through the mere presence of social allies (Sapolsky 1992; Levine 1993a, 1993b; Sachser *et al.* 1998; Smith *et al.* 1998; von Holst 1998). In particular, the hidden stress reduction through passive social support affects the energetic efficiency of the social life of geese, making it a prime mechanism for structuring the greylag goose and many other complex vertebrate social systems. Social alliances are crucial to survival in a stressful social environment, and female geese, in particular, benefit from such alliances, both behaviourally and hormonally. The reduction of stress when accompanied by social allies may be one of the reasons why reproductively successful females are also more likely to be successful in the following year (Cooke *et al.* 1995; Hemetsberger 2002). In addition, social support can explain hitherto inexplicable phenomena such as the adoption of unrelated young (Kalmbach *et al.*

2005; Kalmbach 2006), long-term family formation (Lorenz 1988), and the female-centred clan structure (Frigerio *et al.* 2001a; Chapter 6) at the mechanistic level.

To sum up, being able to survive in a harsh social environment is very probably dependent upon a good social embedding of individuals, based on solid partnerships between mates and the presence of social allies. The stress-reducing effects gained from these are essential in coping effectively with stress, and passive social support, in particular, may affect the energetic efficiency of social life, making it a prime candidate for structuring other vertebrate social systems. We still lack data from several vertebrate groups, particularly social reptiles and fish, but social support may affect individuals of all vertebrate social systems, as we already have examples from mammals and birds.

SUMMARY

Greylag geese show mammal-like active and passive social support both in behaviour and physiology. In various studies, we showed that they actively interfere in agonistic encounters on behalf of their social allies. In addition, greylag geese profit from the presence of an ally even if there is no active involvement, as we were able to show that feeding rates increase, and success in agonistic encounters rises, if a supporter is nearby. The presence of a supporter also influences the physiological stress response: the level of immuno-reactive corticosterone metabolites is reduced if a supporter is nearby, relative to when it is not around. We therefore suggest that social support is one of the prime mechanisms for stress reduction throughout all the social vertebrates.

BRIGITTE M. WEIß, CHRISTIAN SCHLOEGL AND
ISABELLA B. R. SCHEIBER

10

How to tell friend from foe: cognition in a complex society

Social life may pose considerable cognitive demands on animals. For instance, knowing other individuals' relationships relative to oneself or to one another may allow individuals to recruit suitable allies or to avoid costly interactions. This, in turn, would free time for other behaviours, such as feeding or resting. Primates have long been the prime example of animals with rich social networks and profound knowledge about social relationships, including the relatedness of various group members, male–female relationships, and who is an alliance partner to whom (Shettleworth 2010). Almost half a century ago, this led to the formulation of the *social intelligence hypothesis* (Jolly 1966; Humphrey 1976), which proposes that living in complex, primate-like social groups has favoured the evolution of advanced cognitive abilities. In recent years, it has been increasingly recognised that other mammals and some birds, most notably corvids and parrots, rival primates in many aspects of social complexity and cognitive abilities. For instance, apes and corvids show high levels of cooperative behaviour (Warneken & Tomasello 2006; Seed *et al.* 2008; Yamamoto *et al.* 2009), make and use tools (Hunt 1996; Bird & Emery 2009; Lonsdorf *et al.* 2009), may be capable of mental time travel (Clayton & Dickinson 1998; Raby *et al.* 2007; Osvath & Osvath 2008) and tactical deception of others (Whiten & Byrne 1988; Bugnyar & Kotrschal 2004), and may take other's perspectives and attribute knowledge to them (Hare *et al.* 2001; Bugnyar & Heinrich 2005; Dally *et al.* 2006; Bugnyar 2011). Geese,

The Social Life of Greylag Geese: Patterns, Mechanisms and Evolutionary Function in an Avian Model System, ed. I. B. R. Scheiber *et al.* Published by Cambridge University Press. © Cambridge University Press 2013.

however, have not been on this list of clever birds so far. In fact, geese are proverbially considered to be 'silly' in several languages. Yet, as the previous chapters have demonstrated, the social life of geese shares several aspects of social complexity with primates and can be highly demanding. To cope most efficiently with the costs of life in a flock, a goose would greatly benefit by being able to tell 'friends' from 'foes'. Hence, in line with the social intelligence hypothesis, the 'silly goose' may actually be quite clever when it comes to (social) cognition.

10.1 KIN RECOGNITION

Kin discrimination through kin recognition is central to the evolution of social behaviour (Komdeur & Hatchwell 1999) and is frequently an effective means of telling which individuals can be expected to act in a friendly manner towards oneself and which are likely not to do so. Indeed, there is plenty of evidence for preferential allocation of care or aid to close kin (see review by Komdeur & Hatchwell 1999). In geese, kin recognition is important even for newly hatched goslings, which need to know whom to follow upon leaving the nest 24 hours after hatching. Also after fledging, recognition of parents and offspring remains crucial, as it allows parents and offspring to feed together without suffering from aggression and to appropriately direct social support towards family members (Chapter 9). The close spatial proximity among adult sisters (Chapter 6) suggests that kin recognition is still relevant in adult life. The maintenance of family bonds throughout the first year, and to some extent into adulthood, along with the frequent occurrence of social support in greylag goose families, provide strong indications that geese indeed recognise and discriminate between kin and non-kin. Furthermore, the precocial development probably favours rapid development of kin recognition in geese.

This is supported by observations of goslings accidentally getting mixed up with those of another family. If such mixing occurs in the first days after hatching, one or more goslings sometimes end up following the wrong parents, whereas such an involuntary 'adoption' of goslings hardly ever happens once the goslings are 1–2 weeks old (Kalmbach et al. 2005; Kalmbach 2006; Black et al. 2007). Parents typically accept such goslings as their own up to an age of about 2 weeks, but aggressively evict older goslings that may try to follow the family. Similarly, goslings imprinted on humans often accidentally follow people other than their foster parent, particularly if the 'wrong' human walks by and the 'correct' one is stationary. Such errors, however, also cease during the second week of life. These observations not only provide further evidence for kin discrimination but also indicate that geese, like many other bird species (Komdeur & Hatchwell 1999 and references therein), associatively learn who is kin and who is not. For a goose, therefore, kin are not necessarily genetically related, but instead can be perceived as imprinted-on individuals (Bateson 1966). We currently do not know whether geese can recognise kin based on genetic cues alone when social cues are absent.

To strengthen the observational data with experimental evidence, we recently used a systematic approach to study kin recognition in geese. For this purpose, we placed a group of goslings (aged 7–12 days), imprinted on three different human foster parents, in a communal pen. The foster parents placed themselves 10 m from the pen and at an equal distance from each other. Upon releasing the goslings from the pen, the three foster parents simultaneously called the goslings using the same standard words ('come, come'), and monitored their approach. Within only 10 seconds, all of the 15 goslings had gone to their 'own' foster parent (Hohnstein 2010). Interestingly, the goslings had already arranged themselves into their respective sibling groups prior to their release. Hence, individual decisions may have been influenced by the siblings, particularly by the first to approach the foster parent. This experiment therefore does not allow us to test whether goslings recognised their foster parents and/or their siblings, but it does indicate that recognition of individual family members may already occur at a very early age.

To determine who is recognised, we conducted experiments in which individual goslings were given a choice between a family member (foster parent or sibling) and a non-family member (a foster parent or gosling of another sibling group) in a traditional two-choice task. For this purpose, focal goslings were placed in the central compartment of a small arena and a family member and non-family member were presented in two compartments adjacent to the focal compartment (Fig. 10.1). The humans and goslings to choose from were either present 'in person' or their calls were played back from speakers. When first tested at the age of 3 days, goslings showed no preference for kin, independent of the test condition (human or gosling, physically present or playback). At the age of 10 days, however, goslings spent significantly more time closer to their foster parent than to another, familiar human, when the humans were physically present; 12 out of 14 goslings greeted their own foster parent rather than the other human in the test arena, and when the foster parents eventually

Sibling   Focal   Non-sibling

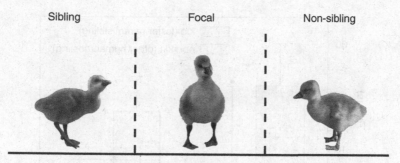

Figure 10.1   Experimental set-up to test kin recognition in greylag goslings. The *hatched line* indicates the wire mesh partitioning the test arena into the focal gosling's central compartment and the two side compartments. The position of the sibling and non-sibling (right or left) was randomised. In other test conditions, the sibling and non-sibling were replaced by the human foster parent and another human or by loudspeakers broadcasting the calls of goslings or humans. Photographs © B. M. Weiß.

left the test arena, 10 of the 12 goslings that attempted to follow the humans followed their own foster parent. Also, the majority of goslings spent more time closer to siblings than to non-siblings, although this difference was not significant (Fig. 10.2). Six focal goslings greeted other goslings during the experiment, and five out of the six greeted their own sibling instead of the non-sibling present (Hohnstein 2010). When kin were not present but were represented by playbacks, goslings did not show a significant preference (A. Hohnstein *et al.*, unpublished). At the age of 10 days, therefore, goslings either were not (yet) able to discriminate kin from non-kin based on acoustic cues alone, or were too distracted when all of their social companions were out of sight. In fact, goslings in the arena frequently paced along the edges in an apparent attempt to find a way out and emitted distress calls, particularly when kin were not visibly present but their calls were played back.

Taken together, the different experiments indicate that goslings first show signs of kin recognition as early as 1 week of age. Orientation towards family members, however, may be absent or faulty under stressful conditions like being physically restrained from approaching the family. During naturally occurring separations from the family, goslings might also rely on responses from their parents, like answering their distress calls or approaching a lost gosling, but we

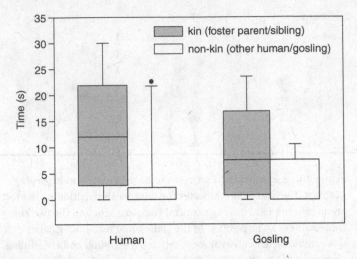

Figure 10.2   Time that focal goslings spent close to kin (*grey bars*) or non-kin (*white bars*) in the kin recognition experiment (see Fig. 10.1) at the age of 10 days. *Boxplots* show medians and quartiles, *whiskers* 10th and 90th percentiles, and *circles* 5th and 95th percentiles.

have not yet investigated experimentally at what age parents reliably distinguish their own goslings from others.

10.2 INDIVIDUAL RECOGNITION

10.2.1 Case studies

It is late February and the mating season is at its peak. After the morning feeding, most of the geese have ventured from the feeding area down to the river bank to rest. Some latecomers are approaching the river bank as the male *LC*, some 500 metres away, raises his head, threatens in the direction of the latecomers, and flies up the river. Most of the latecomers seem to ignore the event, with the exception of the male *Lester*, who escapes by flying away when *LC* is only half way up the river. *LC* gives chase for several hundred metres and then returns to his female partner with a particularly intense triumph ceremony. Similar events happen several times over the course of the day and the following days; the location differs but the two males involved are always *LC* and *Lester*.

At the age of two years, the male *Tarek* forms a pair bond with the female *Judith*. The pair bond is stable until January three years later, when several geese, including *Judith*, get lost during a severe storm and do not return. *Tarek* is subsequently courted by several young males but does not form a tight bond with any of them, nor does he himself court any females for the remainder of the year. In the following spring, he eventually courts and forms a loose bond with a young female. A new student sets out to observe the newly formed pair and after a few days the student reports that *Tarek* is no longer with the young female, but with another goose she was not able to identify. On going out to check, the face of the 'mystery goose' looks quite familiar. She is incompletely ringed but still wears the aluminium ring of the ornithological station Radolfzell. The number confirms that *Tarek*'s new partner is in fact his old one – 15 months after the storm, *Judith* has come home.

10.2.2 Testing true individual recognition

These anecdotes, and the occurrence of stable, long-term pair bonds, suggest that geese – and presumably many other social animals – are perfectly capable of individual recognition. Individual recognition is presumed a prerequisite for complex social interactions and has been described in a variety of species from different taxa. In most studies, however, distinction between individuals could have been achieved through differences in familiarity, or the presence of kin-related or spatial cues. Surprisingly, there is little systematic evidence for true individual recognition in most systems, i.e. recognition based on individually distinct features. One difficulty in testing true individual recognition is that the classical two-way choice experiments used in recognition tests (see the set-up described above) are not applicable to this question, as here focals are expected to respond differently depending upon which individuals are present. However, if the individuals present are equally related, equally familiar, of the same sex etc., one would not predict a difference in behaviour towards either individual. One of the few studies that found a clever way around this problem used an expectancy violation paradigm to determine individual recognition in African elephants (*Loxodonta africana*). Bates and co-workers (2008) placed freshly collected urine of females in the path of travelling female elephants. The elephants reacted differently depending on whether the encountered situation was expected (urine from an animal that was ahead) or unexpected (urine from an animal that should not have been there yet), indicating that the elephants knew the location of other group members in relation to themselves.

As geese, in contrast with most mammals, do not appear to be olfactorily oriented, and family members rarely spread apart, we had to come up with a different way of testing for discrimination between equally related and equally familiar individuals. Rather than making geese choose between different individuals, we made individual identity the cue for what to choose in a classical operant two-choice task. We trained 6-week-old goslings to obtain a piece of bread from one of two cups. The two cups were covered with lids marked with different geometrical symbols and the only cue to the correct choice was the identity of a sibling present during the test trial. For example, the focal individual was presented with cups covered with a triangle and a circle. In the presence of sibling A, the baited cup was covered with the triangle, while the circle was the correct choice in the presence of sibling B (Fig. 10.3). In a first training step, focals were allowed to feed from cup A in the presence of sibling A and from cup B in the presence of sibling B, but subsequently focals had to choose between cups A and B in the presence of only one of their siblings. Within 6 weeks, 10 out of 15 goslings had learned to pick the correct cup at a rate above chance, thereby demonstrating that they distinguished between their two target siblings. The other goslings may have been equally capable of discriminating between their siblings, but may have failed to associate the particular sibling's presence with the choice task within the time frame of the study. Notably, the task itself was based on more or less simple associative learning; however, the successful formation of the associations was conditional on each subject's ability to discriminate between individual siblings. Another aspect is that performance was independent of whether the two target siblings were genetically related or not, or whether they were of the same or different sex. Discrimination, therefore, could neither be based on familiarity, sex nor on genetic kin-related cues, and the geese thus demonstrated true individual recognition based on individually distinct features of each sibling (Scheiber et al. 2011).

It remains to be investigated whether the geese spontaneously discriminated individual features and learned just the association between symbol and individual, or whether they also had to learn to notice the individual features of their siblings. Also, we do not yet know which features the geese used to discriminate between the siblings. In birds, olfaction is thought to be one of the least developed senses (Roper 1999; but see Hagelin & Jones 2007; Steiger et al. 2008; Birkhead 2012), but given the importance of visual, and in particular facial cues, the latter are the prime candidates for future investigations. Indeed, facial cues have been shown to be of importance in discrimination

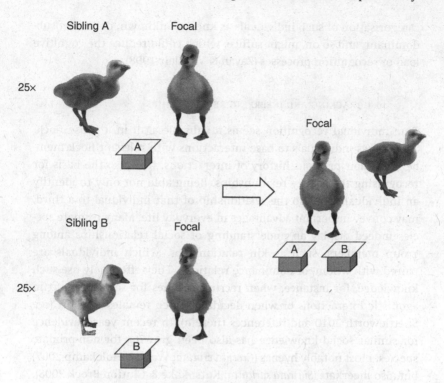

Figure 10.3 Experimental set-up to test discrimination between individual siblings. In a first training step, focals were allowed to feed 25 times from a cup covered with symbol A (e.g. a triangle) in the presence of sibling A and to feed 25 times from a cup covered with symbol B (e.g. a circle) in the presence of sibling B. Thereafter, focals were presented with two cups, covered with symbols A and B in the presence of either sibling A *or* sibling B, whereby only cup A was rewarded when sibling A was present and only cup B when sibling B was present. Photographs © B. M. Weiß.

tasks involving several species, including paper wasps (*Polistes fuscatus*: Tibbetts 2002), crayfish (*Cherax destructor*: van der Velden *et al.* 2008) and primates (*Pan troglodytes, Macaca mulatta*: Parr *et al.* 2000). It also remains to be tested how many of their flock members geese can tell apart, as it seems unlikely that they would be capable of recognising every single individual when flocks comprise thousands of birds, yet other individuals may be 'socially relevant' and worth recognising. Our data show that geese discriminate kin individually, and it seems likely that this applies also to pair partners. Less frequent interaction partners may still be 'socially relevant' and worth recognising, but categorisation of such individuals as known/unknown, dominant/sub-dominant, and so on, might suffice, which could reduce the cognitive load of recognition processes (Zayan & Vauclair 1998).

10.3 TRACKING AND INFERRING RELATIONSHIPS

True individual recognition seems a valuable skill in a goose flock, as it allows individuals to base interactions with relevant flock members on their previous history of interactions. It is also the basis for recognising third-party relationships. Being able not only to identify an individual, but also the relationship of that individual to a third, may convey important advantages in everyday life. Many primate species indeed display an understanding of social relationships among group members, such as kin relationships, which individuals are paired with whom, or dominance relations. They effectively use such knowledge, for instance, when recruiting allies for social support in agonistic interactions or when deciding which females to court (see Shettleworth 2010 and references therein). In recent years, evidence for similar social knowledge has also been growing for non-primate species, most notably hyenas (*Crocuta crocuta*: Watts & Holekamp 2007) but also meerkats (*Suricata suricatta*: Kutsukake & Clutton-Brock 2008), dogs (*Canis lupus familiaris*: Ward *et al.* 2009), birds (Emery *et al.* 2007; Fraser & Bugnyar 2010a, 2010b) and even fish (see below).

A goose might benefit from knowing who belongs to a family or a pair in a variety of situations. For instance, goose D may be dominant over goose H, but subdominant to H's partner B. If D encounters H on a food patch it would not be safe to displace H if B was somewhere nearby, while D could safely chase H if B was busily feeding on another food patch. As we have seen in previous chapters, such scenarios are indeed relevant in everyday goose life, as geese regularly provide social support and chase individuals that previously displaced a family member (Chapter 9). In the latter case, geese selectively attacked a family member's opponent immediately after the interaction between the family member and its opponent (Scheiber *et al.* 2009b; Chapter 9). This may be indicative of third-party recognition but could also merely be due to the opponent ending up close to a high-ranking flock member in the course of its previous interaction, thereby increasing its chances of being the target of an attack. However, heart rate responses to agonistic interactions provide strong evidence that individual geese are indeed sensitive to the rank relationships of their flock members (Wascher *et al.* 2009; Chapter 8). Anecdotal evidence further suggests that geese can keep track of social relationships between flock members independent of the time and location of interactions, e.g. when males selectively attack a former challenger of a pair bond in different locations across weeks and months, or when geese do not only selectively attack a family member's opponent, but the opponent's kin as well.

Knowing 'in advance' whether it is safe to threaten or better to retreat from a conspecific may save both time and energy which, in turn, can be devoted to other behaviours such as feeding. This knowledge can be achieved by tracking one's own dominance relations with flock members or, more efficiently, by tracking relationships between other flock members and inferring unknown relationships from known ones. If, for instance, an animal observes that A is dominant over B and B is dominant over C, the animal could infer that, most probably, A is also dominant over C. This 'transitive inference' (McGonigle & Chalmers 1977; Gillan 1981) could allow animals to assess their own rank in relationship to flock members without having to interact with each flock member 'personally'.

Two extraordinary studies, one in the cichlid *Astatotilapia burtoni* (Grosenick *et al.* 2007) and one in pinyon jays, *Gymnorhinus cyanocephalus* (Paz-y-Miño *et al.* 2004), provided evidence that animals do indeed use transitive inference for assessing their own dominance rank in relation to a conspecific that has been observed interacting with other group members, but never with oneself. The key feature in these studies was that an individual's previous exposure to agonistic interactions could be rigorously controlled; in other words, the individual's own experience in interactions with conspecifics, on the one hand, and which interactions between conspecifics the animal could observe, on the other hand. Unfortunately, previous knowledge of agonistic relationships is not controllable in a natural setting, and applying the set-ups used in the above studies was therefore not feasible in our flock. Transitive inference, however, may also be tested with operant procedures based, for example, on colour or symbol discrimination. Operant transitive inference tests are considerably easier to conduct than social transitive inference tasks and have been performed with a number of species such as chimpanzees, rhesus monkeys, rats *Rattus norvegicus*, pigeons *Columba livia*, and hooded crows *Corvus corone cornix* (Gillan 1981; von Fersen *et al.* 1991; Davis 1992; Treichler & van Tilburg 1996; Lazareva *et al.* 2004). Notably, operant tasks have also been used to test for transitive inference in several species of jays, including the pinyon jay (Bond *et al.* 2003, 2010), where transitive inference skills in both the social and the operant task were shown. Furthermore, the performance of jays and lemurs in operant transitive inference tasks was related to the complexity of the social system (Bond *et al.* 2003; MacLean *et al.* 2008). This suggests that transitive inference may be applied flexibly across contexts, and that operant tasks may be a useful tool to test for transitive inference skills in a social animal when social tasks are not feasible.

Therefore, we used a well-established operant task in which individual geese had to track food rewards in differently coloured cups (Weiß *et al.* 2010a; I. B. R. Scheiber *et al.*, unpublished). The colours were ordered in an arbitrary hierarchical series that acted as a substitute for a dominance hierarchy among flock members. In this task, a goose was presented with two cups and was allowed to open one of them (Fig. 10.4). The correct choice (the cup containing a piece of bread) was indicated by the colour of the lid, whereby colour A was always rewarded when paired with colour B, B was rewarded when paired with C, C when paired with D, and so on. Hence, geese first had to learn to track transitive dyadic relationships, where the correct choice did not depend upon the colour *per se*, but on the combination of colours presented. Importantly, the geese were only presented with adjacent colours in the series. In this set-up, the geese readily learned a five-colour or a seven-colour series, picking the correct colour in each colour pair at a rate above chance, and hence tracked four or six dyadic colour relationships simultaneously. After the training on the adjacent colour pairs was completed, the geese received unrewarded trials with two colours that had never been presented together before; for example colours B and D in a five-colour series or colours C and F in a seven-colour series. They accurately picked the 'higher-ranking' colour well above chance, thereby demonstrating the ability of transitive inference.

Transitive inference has long been assumed to demonstrate an individual's reasoning abilities or, in human terms, 'logical thinking' (McGonigle & Chalmers 1977; Davis 1992). In the last few decades, however, evidence has accumulated that transitive inference problems may be solved by 'simple' associations (Wynne 1997; Van Elzakker *et al.* 2003; Vasconcelos 2008), and that even humans may fall back on an association-based mechanism when not explicitly aware of a logical hierarchy (Frank *et al.* 2005). In fact, the five geese which were trained on a five-colour series (Weiß *et al.* 2010a) may also have had a fairly simple job to do. As colour A was always rewarded and E never was, the only real novel pair was B–D. B, however, was closely associated with A, and D with E, so the geese might have transferred the strong

Figure 10.4   A juvenile goose picks a cup in a two-choice task used to test transitive inference in geese. © S. Kehmeier. A full-colour version is included in the colour plate section.

positive association of A to B and the negative association of E to D (so-called 'value transfer'; e.g. von Fersen *et al.* 1991). Also, the geese had learned the series sequentially forward – from A to E. They might thus have preferred B over D because B was learned earlier ('primacy effect') or because they had had more exposure to the B–C pair during training than to the D–E pair and consequently had formed a stronger reward association for colour B than for D (Vasconcelos 2008). For these reasons, the training method and use of the five-colour series did not allow us to distinguish between such an association-based mechanism and a cognitively more demanding, relational representation, in which some mental representation of the entire series is formed (comparable to a cognitive map; see Tolman 1948; O'Keefe & Nadel 1978). Importantly, the two possibilities do not seem to be mutually exclusive, as recent research suggests that animals may develop both associative and the presumably more complex relational representations of pairs of stimuli (Jacobs 2006; Lazareva & Wasserman 2006). The degree of reliance on one or the other may depend on other factors, such as social complexity or reliance on caching in corvids. By comparing the predicted choice accuracies of a model simulating a purely associative representation with the performance observed in four different species of jays, Bond and co-workers (2010) demonstrated that species living in more complex social systems relied more heavily on relational representations. Along the same lines, MacLean *et al.* (2008) found that the performance of the highly social ring-tailed lemurs (*Lemur catta*) in a transitive inference task was suggestive of a relational representation, while the performance of the less social mongoose lemurs (*Eulemur mongoz*) was better explained by an associative mechanism. Furthermore, the reliance on food caches predicted the reliance on representational mechanisms in corvids independent of the complexity of the social system (Bond *et al.* 2010).

To get an idea about the cognitive mechanisms involved in transitive choice in geese, we repeated the experiment by Weiß *et al.* (2010a) with 12 additional individuals. Importantly, half of these learned the sequence 'forward' and the other half 'backward'. The 'forward' geese thus learned the sequence in the same manner as in the first experimental round; that is, from the always rewarded colour to the never rewarded colour. The 'backward' geese, on the other hand, learned the never rewarded colour first and the always rewarded colour last. If the learning order influenced choice in novel colour pairs, the 'backward' geese should prefer the colours learned earlier, i.e. the lower-ranking colours. All 12 geese, however, significantly chose the higher-ranking colours over the lower-ranking ones irrespective of the training method (I. B. R. Scheiber *et al.*, unpublished), providing clear evidence that correct choice in novel colour pairs was not based on a primacy effect or a stronger association of reward with one or the other colour as a result of the training order.

The 12 geese of the second experimental round also learned a seven-colour series instead of a five-colour one. The seven-colour series contains six novel pairs (B–D, B–E, B–F, C–E, C–F and D–F) which do not include one of the 'anchors' (the first, always rewarded, and the last, never rewarded, colour), while the five-colour series only contains one such pair (B–D). Different cognitive mechanisms lead to a number of predictions about performance in these six pairs, because they differ in the distance (i.e. number of interspersed colours) between each other ('symbolic distance effect': e.g. von Fersen *et al.* 1991) and to the 'anchors' ('serial position effect' and 'value transfer': e.g. von Fersen *et al.* 1991; Vasconcelos 2008). The tested geese performed equally well with all novel pairs (I. B. R. Scheiber *et al.*, unpublished) and therefore did not show either a symbolic distance effect or a serial position effect in pairs with the same symbolic distance (B–D, C–E and D–F). The lack of difference between these pairs also provides convincing evidence against value transfer as the underlying mechanism, as value transfer would have predicted a significantly poorer performance or even failure to choose transitively in the C–E pair, which has the greatest distance from the anchors. The interpretations of the symbolic distance effect (or the lack thereof) differ somewhat depending on the different theoretical models, but essentially the results for the geese trained on a seven-colour series provide sound evidence against a mainly associative representation and some evidence for a relational representation. Geese thus seem to solve transitive inference problems mostly based on some form of relational representation and, given their rich social networks, thereby corroborate the findings of Bond *et al.* (2010) about the relationship between cognitive mechanisms and social complexity in corvids. Furthermore, geese raised in smaller sibling groups learned to track dyadic relationships faster than those from larger sibling groups (Weiß *et al.* 2010a). As smaller families are typically involved in more agonistic interactions than large ones (Scheiber *et al.* 2005a), goslings from smaller sibling groups may have had more opportunities to learn about dyadic relationships in the flock. Learning to track dyadic relationships might also have been facilitated in smaller groups with fewer dyads. For substantiated conclusions, however, we will need to better understand the relationship between the early social environment and cognitive abilities in later life. Nonetheless, these results support the idea that different social environments may not only account for between- but also for within-species differences in cognitive abilities.

At 2, 6 and 12 months after the end of the transitive inference tests, we presented the coloured cups to the geese once more. Despite having had no more exposure to the cups and colours during all this time, the geese continued to choose the correct colours from the trained pairs even after half a year and irrespective of whether the geese had learned a five-colour or a seven-colour series (Weiß & Scheiber 2013). Hence, similar to *Tarek* remembering his beloved *Judith* for many months, the geese here demonstrated considerable long-term memory for relationships.

10.4 CONSPECIFICS AS SOURCES OF INFORMATION

Animals may not only need to know something *about others*, but may benefit if they can make use of *what others know*. Ravens and other corvids, for instance, cache and pilfer food (Heinrich & Pepper 1998;

Bugnyar & Kotrschal 2002; Dally *et al.* 2006). Cachers pay close attention to who has watched a cache being made and employ cache protection strategies depending on whether competitors are knowledgeable or naive about a cache location. These findings are highly suggestive of some understanding about what others know (Bugnyar 2011; see also Hare *et al.* 2000 for similar findings in chimpanzees), although compelling evidence is still lacking that non-human animals understand anything about the unobservable mental states of other animals (Penn & Povinelli 2007). Conspecifics, however, are useful sources of information even without any understanding of their mental state. Large aggregations of conspecifics are usually a good indicator of where to find food or safe resting sites, and paying attention to others' behaviours may provide essential information about resources and dangers, but also about social interactions and relationships. This can be achieved with simple cognitive mechanisms: local enhancement (i.e. drawing an observer's attention towards a location where another individual is acting) and stimulus enhancement (i.e. drawing an observer's attention towards a type of object that another is manipulating) (Giraldeau 1997). Fritz *et al.* (2000) showed that stimulus enhancement indeed allowed juvenile greylag geese to exploit food sources not available to other geese. To obtain a food reward, hand-raised juveniles had to open the sliding lid of a box, where some of the individuals had observed the action being performed by a human tutor and others had not. The seven geese that had observed the opening of the box explored more often at the position shown by the tutor than elsewhere and all learned to open the box, seemingly by trial and error. In contrast, the seven control birds explored primarily at positions that did not allow them to open the box and only one of them succeeded. Observing an experienced model was probably also the reason for the rapid spread of a novel foraging tradition in our flock in the late 1990s, when a few individuals bit through the stems of butterbur leaves (*Petasites hybridus*) to chew the ends. By the end of the following summer, this behaviour had spread to almost all individuals of the flock, and had appeared to spread particularly fast in families in which one member was already chewing the plant stems (Fritz *et al.* 1999).

An individual's gaze direction can also convey information about the environment and about an individual's interests and goals (Baron-Cohen 1995). Therefore the ability to detect and to follow the gaze direction of another individual (i.e. 'gaze following': Emery 2000) may be highly beneficial, as it may alert observers to the presence of predators, conspecifics, food or other resources.

Visual co-orientation to a position above, behind or to the side of another individual ('gaze following into distant space': Schloegl *et al.* 2007) has been suggested to be an automatic visual orientation reflex (Povinelli & Eddy 1996). It seems to be a fairly basic cognitive skill that can be found in reptiles (Wilkinson *et al.* 2010), birds (Schloegl *et al.* 2008; Loretto *et al.* 2010) and mammals (Kaminski *et al.* 2005). In corvids, gaze following into distant space emerges shortly after fledging, when birds increase their mobility and become increasingly vulnerable to predators. This observation led to the suggestion that gaze following into distant space may serve as an anti-predator behaviour (Bugnyar *et al.* 2004; Schloegl *et al.* 2007). As such, gaze following would also be highly adaptive in geese and, in contrast to altricial corvids, would be particularly beneficial in very young precocial goslings long before fledging.

We therefore tested greylag goslings for their ability to follow a sibling's gaze into distant space. For this purpose we placed two goslings into an arena which was separated by a partition constructed of wire mesh below and an opaque pane above. This ensured that goslings could see each other through the wire mesh but could not see what was happening on the other side of the opaque partition above the sibling's head (Kehmeier *et al.* 2011). In one of the two goslings to be tested, we elicited a look-up towards the opaque part of the partition by projecting a laser point onto it. We scored whether the sibling on the other side looked up as well. Several control conditions made sure that the sibling's response was not a look up by chance or that the gosling could somehow see the laser point projected on the other side. Goslings were significantly more likely to look up in response to a sibling's look-up than in either control condition; that is, they demonstrated gaze following. Importantly, they already did so when the experiment started at an age of 10 days (Kehmeier *et al.* 2011). This confirmed our prediction that this skill emerges considerably sooner in the precocial greylag geese than in altricial birds, and emphasises the adaptiveness of gaze following.

SUMMARY

The so-called 'silly goose' actually turned out to be quite clever, at least in the social domain. Anecdotal and systematic observations, as well as experiments, show that geese are capable of telling kin from non-kin and one individual from another, and that they can track dyadic relationships and infer unknown relationships from known ones. Furthermore, they can memorise individuals and relationships for months, or even years, and can use their conspecifics as a source of information. In other words, they possess a suite of cognitive abilities that probably help them to manage their complex social life, thereby reversing their 'bad reputation'.

Part IV   Lessons for vertebrate social life

ISABELLA B. R. SCHEIBER, KURT
KOTRSCHAL AND BRIGITTE M. WEIß

11

The greylag goose as a model for vertebrate social complexity

In the Preface we mentioned how Jane Goodall, Irene M. Pepperberg, Sue Savage-Rumbaugh and Konrad Lorenz paved the way for accepting that in some instances there may be emotional involvement in one's study of animals without compromising scientific objectivity. This notion opened new avenues when, for example, investigating the emotions or cognitive skills of animals (Panksepp 1998; Heinrich 1999; Pepperberg 1999; Aureli & Schaffner 2002; Aureli & Whiten 2003; Chapter 10). It is through trusting relationships between humans and focal animals that close and detailed observations by the human observer and one-to-one work with the study animals are possible. This is what makes study systems such as the Gombe chimpanzees or the KLF greylag geese so valuable. Indeed, the complete life history data of the KLF geese collected since 1973 (see Chapter 1) are one of the biggest assets of this flock, providing a unique research opportunity on the social system of a long-lived vertebrate. This final chapter places our findings into an evolutionary framework, and ties them together with findings from other socially complex mammals and birds. Particular focus is given to the 'social brain', addressing the question: 'How large a brain does one need to be able to cope with a complex social environment?'

11.1 THE EVOLUTION OF SOCIALITY

Much interest in studying social behaviour and the resulting social systems was triggered by the contributions of W. D. Hamilton (1964a,

The Social Life of Greylag Geese: Patterns, Mechanisms and Evolutionary Function in an Avian Model System, ed. I. B. R. Scheiber *et al.* Published by Cambridge University Press. © Cambridge University Press 2013.

1964b), Jerram L. Brown (1975) and E. O. Wilson (1975a), among others, who attempted to explain social behaviour as the outcome of natural selection *sensu* Darwin (1859). In her comprehensive analysis of chimpanzee behaviour, Jane Goodall (1986) stressed that chimpanzees do not only show definite personalities but also a social system which is quite complex and remarkably similar to those of human societies. As humans and chimpanzees are closely related phylogenetically, it would actually be more surprising if this were not the case. Birds and mammals, however, are phylogenetically distant. Common patterns of social complexity found in geese and primates, in their social dispositions and behaviour, are·in this case very unlikely to be simply due to common descent. The potential to deduce general principles of complex social behaviour thus may be greater when comparing distantly related species, such as birds and primates, and more informative in explaining general social structures.

Indeed, analogous social structures can be found in taxa that split long ago; for instance 280–300 million years in the case of mammals and birds or more than 400 million years in mammals and fish (Jarvis *et al.* 2005), and sociality arose independently at various times in very diverse lineages throughout the vertebrates. Nevertheless, a variety of structures and mechanisms that govern sociality are highly conserved throughout the vertebrates. These include the homologous 'social behaviour network' within the basal forebrain and midbrain in mammals, birds and fish (Goodson 2005), a bonding reward mechanism in the brain involving two closely related neuropeptides: oxytocin (mesotocin being the avian homologue) and arginine vasopressin (Curley & Keverne 2005), basic emotional systems (Panksepp 1998), and the fast and slow stress coping systems: the sympatho-adrenomedullary (SAS) and hypothalamic–pituitary–adrenocortical (HPA) stress axes (Sapolsky 1992; von Holst 1998; Summers 2002). Social contexts were found to be among the greatest modulators of these systems (DeVries *et al.* 2003; McEwen & Wingfield 2003; Goymann & Wingfield 2004; Chapters 8 and 9).

There is no general consensus as to how to define a complex social system (Kotrschal *et al.* 2010). Some authors might consider the most advanced form of sociality to be eusociality, characterised by overlapping generations, cooperative care of the young, and reproductive division of labour, which includes sterility or near-sterility of the overwhelming majority of colony members. The latter forego direct reproduction, at least temporarily, and help others to reproduce (Korb 2010). As most practitioners of eusociality are insects, invertebrate biologists tend to favour this definition of social complexity.

Complex social systems in vertebrates, however, are more often defined through the variety of relationships and interactions that members of a group maintain between themselves. In particular, the preservation of long-term, mutually valuable relationships among mates, parents and offspring, or unrelated social allies (van Schaik & Aureli 2000), the varied patterns of conflict resolution (Aureli & de Waal 2000; Aureli *et al.* 2002), the ability to track multiple-level relationships (Judge & Mullen 2005), and social support (Sapolsky 1993; Goymann & Wingfield 2004; Chapter 9) describe social complexity. Few studies so far have investigated avian social relationships in the same way as those of primates (Iwaniuk and Arnold 2004), but recent evidence suggests that birds and other vertebrates do not fall short in terms of social complexity. Corvids, in particular, were studied extensively in this respect (Emery *et al.* 2007; Fraser & Bugnyar 2010a, 2010b). Our studies on the greylag goose social system discussed in the preceding chapters may also have a contribution to make.

11.2 DO BIRDS HAVE THE BRAINS FOR SOCIAL COMPLEXITY?

Managing complex social relationships may require specific cognitive adaptations. These days, research seems to converge on the generalisation that cognitive evolution is well explained by responses to challenges from the social environment (Cheney & Seyfarth 1986; Byrne & Whiten 1988; but see Whiten & Byrne 1997; van Schaik & Deaner 2003; Barrett *et al.* 2007; Moll & Tomasello 2007). Similarly, social behaviour has frequently been cited as a significant factor in the evolution of brain size and structure in birds and mammals (Wyles *et al.* 1983; Gittleman 1986; Byrne & Whiten 1988; Dunbar 1992, 1995; Marino 1996; Dunbar & Bever 1998; Delius *et al.* 2001; Iwaniuk & Nelson 2003; Iwaniuk & Arnold 2004; Shultz & Dunbar 2006; but see Beauchamp & Fernandez-Juricic 2004; Healy & Rowe 2007; Holekamp 2007). For a long time, birds have been believed to be incapable of complex cognition because of their mainly striatal forebrain. It was thought that birds were mostly capable of instinctive behaviour because the avian cerebrum (telencephalon) was thought to be composed of almost entirely basal ganglia. Today we know that this is, in fact, incorrect: some previously assumed 'primitive' avian pallial areas, i.e. the hyperpallium, the nidopallium, and the arcopallium, are indeed processing regions performing functions similar to those of the mammalian cortex and are homologous to certain mammalian pallial regions (Jarvis *et al.* 2005).

Recently, Emery *et al.* (2007) reported on cognitive adaptations of social bonding in birds. They postulated that lifelong pair bonds, monogamy and complex interactions in the pair may be the reason for a relatively large brain size in birds, whereas in mammals large brains are the result of group interactions (but see Shultz & Dunbar 2007). The authors postulate 'some form of relationship intelligence' and claim that the quality of the relationship between pair-bonded individuals in greylag geese is not as sophisticated as that shown in rooks (*Corvus frugilegus*) and Eurasian jackdaws (*C. monedula*), due to their relatively smaller brain. This led us to scrutinise this statement (Scheiber *et al.* 2008).

The first question is whether greylag geese are indeed 'small-brained'. Iwaniuk and Hurd (2005) demonstrated that, similar to mammals, 67 avian species could be sorted into five defined cerebrotypes, termed clusters A–E. Cerebrotypes are a species-by-species measure of brain composition that reflect specific ecological niches and/or phylogenetic clades. Iwaniuk and Hurd's (2005) cluster D, which possesses the largest nidopallial, mesopallial and striatopallial volumes of any cluster, indicates that here the telencephalon occupies a much larger proportion of the brain. Cluster D is composed of parrots, passerines, including the carrion crow (*C. corone*), and – surprisingly – waterfowl, including the greylag goose (Iwaniuk & Hurd 2005). The authors note that both parrots and passerines, which are known to perform more complex cognitive tasks than most other bird species, possess relatively large telencephalic volumes (Burish *et al.* 2004; Iwaniuk & Hurd 2005; Iwaniuk *et al.* 2005), and that these may be related to complex cognitive behaviour such as social interactions (Burish *et al.* 2004; Iwaniuk *et al.* 2005). The nidopallium, for example, is the major site of information integration in the avian brain, the mesopallium is involved in a variety of cognitive tasks including social behaviour (Ball & Balthazar 2001), and the striatopallium is essential in learning. Iwaniuk and Hurd (2005) therefore consider cluster D to be a 'cognitive' cerebrotype. This strongly suggests that greylag geese possess the necessary hardware to use advanced cognitive abilities.

Regardless of whether greylag geese are relatively large- or small-brained, this raises the question of whether relative brain size in birds is a valid proxy for 'social intelligence' when comparisons with distant taxa are performed. Developmental modes in birds, for example, influence relative brain size. Birds with altricial development possess relatively larger brains than birds with precocial development (Nealen & Ricklefs 2001; Iwaniuk & Nelson 2003; Burish *et al.* 2004; Iwaniuk & Hurd 2005; Isler & van Schaik 2009). The terms 'altricial' and 'precocial' refer to the extremes of a spectrum of increasing maturity at hatching and decreasing dependence on parental care (Gill 1994). However, if one only considers the Anatidae, a within-clade comparison reveals that greylag geese are among the 'brainiest' in this family (Ebinger & Löhmer 1987).

Furthermore, cooperative breeding as one aspect of avian sociality is present in many or most members of the fairy-wren, honeyeater, woodswallow and crow families (Maluridae, Meliphagidae, Artamidae, Corvidae). It was suggested to be a derived trait in the passeriform parvorder Corvida (Cockburn 1996). Iwaniuk and Arnold (2004) investigated whether the evolution of cooperative breeding is correlated with relatively large brains in 155 species within this parvorder. Most comparisons in this study, however, revealed insignificant results regardless of whether phylogenetic relationships were taken into account or not; i.e. they found no correlation between relative brain size and the evolution of cooperative breeding.

Finally, domestication is known to have a significant effect on relative brain size and brain composition (Iwaniuk & Hurd 2005). Relative to greylag geese, the largest reduction in brain size in domestic geese (*Anser anser domesticus*), i.e. 31%, is found in the hippocampus, located in the avian pallium. The avian hippocampus is considered to be homologous to the mammalian hippocampus, and in both cases is involved in an overall ability for spatial learning and memory (Macphail 2001). Emery *et al.* (2007) speculated on two neural systems relating to social behaviour in monogamous bird species which are known to flock outside the breeding seasons: a *'core neural system'*, which may function as a lasting storage of information about the bonding partner, representing both past events and declarative knowledge, and possibly located in the hippocampus, as well as a more plastic *'secondary neural system'*, which would function in processing and retaining information about transient social relationships outside the breeding season (Emery *et al.* 2007), located in the caudal nidopallium. They suggest that the social memories in the nidopallium may be recent and would have to be updated continuously, whereas in the hippocampus past memories about a partner's response with regard to the social environment are stored in a long-term manner (Emery *et al.* 2007). Interestingly, these two brain regions once more stand out in greylag geese, hinting again at a possible connection with 'social intelligence'.

11.3 SOCIALLY COMPLEX GEESE?

Although, until recently, most researchers interested in social behaviour would have been amused by the notion that greylag geese might be playing in the same league as mammals and corvids in terms of their social complexity (Emery *et al.* 2007), our recent contribution concerning 'Individual performance in complex social systems: the greylag goose example' (Kotrschal *et al.* 2010) has convinced others that this may be indeed the case. Kutsukake (2011) states in his review of the recently published book entitled *Animal Behaviour: Evolution and Mechanisms* (Kappeler 2010):

> Of interest is that this species demonstrates social traits that partially resemble some of those exhibited by Old World monkeys (macaques and baboons), such as cohesive social groups based on female-centred clans, female philopatry, long-term parent–offspring association, and mutual support among related females. This species also shows sophisticated socio-cognitive abilities such as understanding transitive inferences. These characteristics fit the criteria used by primate researchers for 'social complexity'.

So, why are there still sceptics to this claim? The reason may be that supposedly the bonds of greylags are not as elaborate (Emery *et al.* 2007), because they lack certain behaviours, notably allo-preening, which is considered to be necessary for complex social bonds. But is this really the case?

Valuable relationships are characterised by close proximity between bonded partners, the provision of social support, low rates of aggression, and the occurrence of affiliative behaviours (van Schaik & Aureli 2000). Affiliative behaviours, which are thought to provide the basis for solid partnerships in mammals, include allo-grooming (its avian equivalent being allo-preening: Brown 1987), allo-feeding and food sharing (Aureli & de Waal 2000) as well as greeting of the partner(s) (Smith *et al.* 2011). These behaviours are displayed to varying degrees in a variety of bird species (Senar 1984; Brown *et al.* 1991; Zann 1996; de Kort *et al.* 2003, 2006). Although greylag geese never allo-preen or allo-feed, food sharing and partner greeting do occur (Black & Owen 1989a; Lorenz 1991). Does this now imply that the lack of some of these behaviours in geese results in weaker social bonds?

We suggest that the difference in displays of social bonds of geese and corvids is based on their different developmental modes: corvids are altricial, geese are precocial. Allo-grooming and allo-feeding performed between bonded partners in altricial birds is presumably derived from behaviours administered by parents to their hatchlings as a result of their developmental mode (Starck & Ricklefs 1998). Altricial development much more closely resembles the development of most mammalian species: the newly hatched or born offspring are relatively immobile and must be cared for by adults. Altricial birds need to be fed for extended periods of time after hatching, and normal feeding behaviour requires the interaction of chick and parent through a variety of auditory, visual and tactile cues. Nestlings usually receive food by direct insertion into the beak or regurgitation (Gill 1994). It should be mentioned that parental feeding alone does not allow a discrimination between precocial and altricial developmental modes, as in a very few precocial species, for example magpie geese, *Anseranas semipalmata*, some species of loons (Gaviidae), grebes (Podicipedidae), and grouse (Tetraonidae) parental birds also feed their young (Starck & Ricklefs 1998). Another parental duty in altricial species in which parents touch their young is to rid them of ectoparasites so as to prevent the nest and nestlings from becoming a breeding ground for disease (Gill 1994). This is also not necessary in precocial species. The above-mentioned tactile behaviours and the necessity for extended physical contact are not part of the behavioural repertoire of greylag geese. In fact, the only times when individual geese have tactile contact is (i) when they copulate, (ii) during severe aggressive interactions between males, and (iii) when females brood hatchlings during their first few weeks of life (Fig. 11.1). Consequently, it should not be surprising that greylag geese lack behaviours such as allo-grooming and allo-feeding, which express strong social bonds in corvids and other altricial birds. Partnerships in geese are displayed exclusively in non-tactile interactions (Lorenz 1991) and, analogous to the expression of emotions via facial mimics of primates, geese make use of specific neck and body postures (Fischer 1973a, 1973b) in the manifestation of their emotional states. Contrary to altricial birds, in which, in addition to close proximity, solid partnerships are displayed by various affiliative behaviours, precocial geese display and reinforce their social bonds by being close to their partners (Lorenz 1991; Frigerio *et al.* 2001a) and by performing the 'triumph ceremony' together (Fischer 1965; Radesäter 1974). The triumph ceremony consists of a series of ritualised movements accompanied by an impressive vocal display, with the last element being a concerted cackling of male and female (Fischer 1965). The triumph ceremony only occurs in pairs, the elementary social unit of the flock, or within families ('*Triumphgemeinschaft*': Fischer 1965) and is similar to certain behaviours displayed by altricial birds including corvids (common ravens, *C. corax*: '*dickköpfiges Imponieren*', '*Verbeugungszeremonie*': Gwinner 1964; 'self-aggrandizing display': Heinrich 1989; jackdaws: C. Schwab, personal communication), although in corvids these behaviours are displayed in other contexts as well.

The final point to consider is whether proximity between social allies is a fair proxy for the strength of a social bond. In most studies

Figure 11.1   The female, *Lestate*, brooding goslings. Greylag goose families display close parent–offspring bonds, which include tactile contact in the first few weeks after hatching, as goslings cannot keep their body temperature constant. After that time, family bonds are displayed solely in a non-tactile way through close proximity and certain behaviours, including the triumph ceremony. © B. M. Weiß. A full-colour version is included in the colour plate section.

investigating attachment, proximity between social allies has been used as a measure (Lamprecht 1977; Spoon *et al.* 2004) or it was postulated that spatial association can be used to assess socio-positive relationships between individuals (van Schaik & Aureli 2000; Bonnie & de Waal 2006). The utility of using proximity as an indicator of close social bonds was assumed also in several studies on social networks (Croft *et al.* 2004, 2005, 2006; Tóth *et al.* 2009; Gygax *et al.* 2010). A recent study, which specifically asked whether the main characteristics of nearest neighbour (i.e. distance) and socio-positive (i.e. allo-preening, contact sitting) networks are congruent, showed that, indeed, proximity may be used to make certain statements with regard to affiliations between two individuals if sample sizes are large enough (C. Schwab, personal communication).

   One important advantage of close proximity between pair partners is the ability of a partner to provide protection; and an increase in stress can be measured if the social partner is not nearby (see Chapter 8). Young animals, in particular, are highly motivated to form attachments for their own safety, and respond strongly if they perceive themselves to be at risk when their attachment figure is not in close proximity (Cords & Aureli 2000). This distress of being separated from a partner, or the relief and comfort experienced when reunited, can also be measured in adults, both behaviourally and physiologically (Boissy & LeNeindre 1997; Christenfeld *et al.* 1997; Ruis *et al.* 2001; Remage-Healey *et al.* 2003), and greylag geese are no exception (I. Nedelcu *et al.*, unpublished). During separation, greylag geese utter contact calls, and during reunions after separation greylag geese 'greet' one another. This display has been described as higher-intensity contact calling, with the neck stretched forward, and is only performed between socially bonded geese (Lorenz 1991), after which the reunited geese stay close to one another. Along the same lines, geese seek the close proximity of their partners after unsuccessful agonistic interactions (O. N. Fraser *et al.*, unpublished).

In summary, despite differences in how social bonds are displayed in geese, they are certainly comparable with those shown by other birds and mammals. The lack of commonly recognised affiliative behaviours which are used as a measure of the strength of a pair bond in altricial birds and mammals does not imply weaker social bonds; instead these behaviours are not part of the goose's behavioural repertoire due to their different developmental mode. Other behaviours, such as the triumph ceremony, the 'greeting' of social allies, behavioural and hormonal synchronisation between successful pairs (see Chapter 4), and their close proximity are as appropriate in determining who is socially bonded to whom.

11.4 THE GREYLAG GOOSE AS A MODEL SYSTEM FOR SOCIAL COMPLEXITY IN VERTEBRATES

Finally, what can the geese teach us about the evolution of complex vertebrate social systems in general? Are they indeed a suitable vertebrate model for the study of social complexity? One major conclusion that we have drawn from our work is that geese exhibit parallels with social mammals in many ways. These parallels in social complexity and social cognition suggest that these features are probably much more widespread throughout the vertebrates than previously assumed. On the other hand, there are certain rules that have to be followed when living socially, including effective stress management through social support or the notion that the expression of individual dispositions ('personality') may need to be suppressed in the social context. All social vertebrates, including primates, dolphins and corvids, may face similar challenges both from their social and non-social environment, and they have to cope with them efficiently. Bearing this in mind, research on social systems should be broadened, including not only the classical 'socially intelligent' species but also any other social vertebrate system. Including species of varying 'cleverness' (i.e. differing relative brain sizes) would allow an assessment of how much 'brain power' is indeed necessary to evolve social complexity. The similarities and differences between various species will allow us to understand the evolution of vertebrate social systems as a whole.

In closing, we found that the greylag geese at the KLF proved to be a suitable model for looking at social complexity. This semi-tame, yet socially intact, flock allowed us to examine various aspects of sociality over several decades under natural conditions as well as when being experimentally manipulated. The long-term data from this flock include the social relationships between all individuals, their dominance rank, reproductive success and other life history data. As Clutton-Brock and Sheldon (2010) have noted, these long-term studies are most valuable and once abandoned cannot be revived easily, if at all. As long as our geese cooperate, there are many more lines of research to be explored.

SUMMARY

In this final chapter, we discuss whether the greylag goose social system can be considered a model system for other social vertebrates as well. We argue that the social system of greylag geese is complex and comparable to those of many mammalian species as well as the more 'brainy' birds such as corvids and parrots. We find various types of relationships and interactions between flock members, including mutually valuable relationships *sensu* van Schaik and Aureli (2000), various forms of conflict affiliation, and social support between socially bonded individuals, which resemble those of a variety of mammals. The enlargement of certain areas of the brain (i.e. the telencephalon) indicates that greylag geese may have the tools to track and trace a large amount of social information, similar to corvids and parrots, and they have also evolved specific bonding behaviours that are displayed only between individuals that maintain valuable relationships. The fact that bonding behaviours such as allo-preening are lacking in geese may be rooted in differences in their developmental mode: whereas geese are precocial, corvids, parrots and most mammals are altricial. When bearing this in mind, we conclude that greylag geese are, indeed, a suitable model for studying vertebrate social complexity, and that the KLF flock provides a valuable resource for doing so.

References

Abbott D. H., Keverne E. B., Bercovitch F. B., *et al.* (2003). Are subordinates always stressed? A comparative analysis of rank differences in cortisol levels among primates. *Hormones and Behavior*, 43: 67–82.

Åhlund M. (2005). Behavioural tactics at nest visits differ between parasites and hosts in a brood-parasitic duck. *Animal Behaviour*, 70: 433–40.

Åhlund M., Andersson M. (2001). Brood parasitism: female ducks can double their reproduction. *Nature*, 414: 600–1.

Aicher P. J. (2001). *Rome Alive: A Source-Guide to the Ancient City*, Vol. 1. Bolchazy-Carducci Publishers, Mundelein, IL.

Albon S. D., Staines H. J., Guinness F. E., *et al.* (1992). Density-dependent changes in the spacing behaviour of female kin in red deer. *Journal of Animal Ecology*, 61: 131–7.

Alexander R. D. (1974). The evolution of social behavior. *Annual Review of Ecology and Systematics*, 5: 325–83.

Anderholm S., Waldeck P., van der Jeugd H. P., *et al.* (2009a). Colony kin structure and host–parasite relatedness in the barnacle goose. *Molecular Ecology*, 18: 4955–63.

Anderholm S., Marshall R. C., van der Jeugd H. P., *et al.* (2009b). Nest parasitism in the barnacle goose: evidence from protein fingerprinting and microsatellites. *Animal Behaviour*, 78: 167–74.

Angelier F., Weimerskirch H., Dano S., *et al.* (2007). Age, experience and reproductive performance in a long-lived bird: a hormonal perspective. *Behavioral Ecology and Sociobiology*, 61: 611–21.

Anzenberger A. A. (1983). *Bindungsmechanismen in Familiengruppen von Weissbüscheläffchen (Callithrix jacchus)*. Dissertation, Philosophical Faculty, University of Zurich.

Arcese P., Smith J. N. M. (1985). Phenotypic correlates and ecological consequences of dominance in song sparrows. *Journal of Animal Ecology*, 54: 817–30.

Arnold W., Dittami J. P. (1997). Reproductive suppression in male alpine marmots. *Animal Behaviour*, 53: 53–66.

Arnold W., Ruf T., Reimoser S., *et al.* (2004). Nocturnal hypometabolism as an overwintering strategy of red deer (*Cervus elaphus*). *American Journal of Physiology*, 286: 174–81.

Arnqvist G., Rowe L. (2005). *Sexual Conflict*. Princeton University Press, Princeton, NJ.

Aureli F., de Waal F. B. M. (2000). *Natural Conflict Resolution*. University of California Press, Berkeley, CA.

Aureli F., Schaffner C. M. (2002). Relationship assessment through emotional mediation. *Behaviour*, 139: 393–420.

Aureli F., van Schaik C. P. (1991). Post-conflict behaviour in long-tailed macaques (*Macaca fascicularis*). II. Coping with the uncertainty. *Ethology*, 89: 101–14.

Aureli F., Whiten A. (2003). Emotions and behavioral flexibility. *In* Maestripieri D. (ed.) *Primate Psychology: The Mind and Behavior of Human and Nonhuman Primates*. Harvard University Press, Cambridge, MA, pp. 289–323.

Aureli F., Veenema H. C., Van Panthaleon Van Eck C. J., *et al.* (1993). Reconciliation, consolation, and redirection in Japanese macaques (*Macaca fuscata*). *Behaviour*, 124: 1–21.

Aureli F., Preston S. D., De Waal F. B. M. (1999). Heart rate responses to social interactions in freemoving rhesus macaques: a pilot study. *Journal of Comparative Psychology*, 113: 59–65.

Aureli F., Cords M., van Schaik C. P. (2002). Conflict resolution following aggression in gregarious animals: a predictive framework. *Animal Behaviour*, 64: 325–43.

Austin G. E., Rehfisch M. M., Allan J. R., *et al.* (2007). Population size and differential population growth of introduced Greater Canada Geese *Branta canadensis* and re-established Greylag Geese *Anser anser* across habitats in Great Britain in the year 2000. *Bird Study*, 54: 343–52.

Axelrod J., Reisine T. D. (1984). Stress hormones: their interaction and regulation. *Science*, 224: 452–9.

Ball G. F., Balthazar J. (2001). Ethological concepts revisited: immediate early gene induction in response to sexual stimuli in birds. *Brain, Behavior and Evolution*, 57: 252–70.

Balshine S., Kempenaers B., Szekely T. (2002). Conflict and cooperation in parental care: introduction. *Philosophical Transactions of the Royal Society B*, 357: 237–40.

Baron-Cohen S. (1995). *Mindblindness: An Essay on Autism and Theory of Mind*. MIT Press, Cambridge, MA.

Barrett L., Henzi P., Rendall D. (2007). Social brains, simple minds? Does social complexity really require cognitive complexity? *Philosophical Transactions of the Royal Society B*, 362: 561–75.

Barta Z., Houston A. I., McNamara J. M., *et al.* (2002). Sexual conflict about parental care: the role of reserves. *American Naturalist*, 159: 687–705.

Bartolomucci A., Palanza P., Sacerdote P., *et al.* (2005). Social factors and individual vulnerability to chronic stress exposure. *Neuroscience and Biobehavioral Reviews*, 29: 67–81.

Barton R. A., Whiten A. (2002). Feeding competition among female olive baboons, *Papio anubis*. *Animal Behaviour*, 46: 777–89.

Bates L. A., Sayialel K. N., Njiraini N. W., *et al.* (2008). African elephants have expectations about the locations of out-of-sight family members. *Biology Letters*, 4: 34–6.

Bateson P. (1966). The characteristics and context of imprinting. *Biological Reviews*, 41: 177–316.

Bauer H.-G., Bezzel E., Fiedler W. (2005). *Kompendium der Vögel Mitteleuropas*, Band 1. AULA-Verlag, Wiebelsheim, Germany.

Baugh A. T., Schaper S. V., Hau M., *et al.* (2012). Corticosterone responses differ between lines of great tits (*Parus major*) selected for divergent personalities. *General and Comparative Endocrinology*, 175: 488–94.

Beaman M., Madge S. (1998). *Handbuch der Vogelbestimmung: Europa und Westpaläarktis*. Eugen Ulmer GmbH & Co., Stuttgart, Germany.

Beauchamp G., Fernandez-Juricic E. (2004). Is there a relationship between fore-brain size and group size in birds? *Evolutionary Ecology Research*, 6: 833–42.

Begin J., Beaugrand J. P., Zayan R. (1996). Selecting dominants and subordinates at conflict outcome can confound the effects of prior dominance or subor-dination experience. *Behavioural Processes*, 36: 219–26.

Bell A. M. (2007). Future directions in behavioural syndrome research. *Proceedings of the Royal Society B*, 274: 755–61.

Bell A. M., Hankison S. J., Laskowski K. L. (2009). The repeatability of behav-iour: a meta-analysis. *Animal Behaviour*, 77: 771–83.

Bem D. J., Allen A. (1974). On predicting some of the people some of the time: the search for cross-cultural consistencies in behavior. *Psychology Reviews*, 81: 506–20.

Benton M. J., Donoghue J. P. (2007). Paleontological evidence to date the tree of life. *Molecular Biology and Evolution*, 24: 26–53.

Benus R. F., Bohus B. G., Koolhaas J. M., *et al.* (1991). Heritable variation for aggression as a reflection of individual coping strategies. *Experientia*, 47: 1008–19.

Bercovitz A., Collins J., Price P., *et al.* (1982). Noninvasive assessment of seasonal hormone profile in captive bald eagles (*Haliaetus leucocephalus*). *Zoo Biology*, 1: 111–17.

Bercovitz A., Czekala N. M., Lasley B. L. (1988). A new method of sex determi-nation in monomorphic birds. *Journal of Zoo Animal Medicine*, 9: 114–24.

Bergmüller R. (2010). Animal personality and behavioural syndromes. In Kappeler P. M. (ed.) *Animal Behaviour: Evolution and Mechanisms*. Springer Verlag, Berlin, pp. 587–621.

Berkman L. F., Leo-Summers L., Horwitz R. I. (1992). Emotional support and survival after myocardial infarction: a prospective, population-based study of the elderly. *Annals of Internal Medicine*, 117: 1003–9.

Berman C. M. (1980). Early agonistic experience and rank acquisition among free-ranging infant Rhesus monkeys. *International Journal of Primatology*, 1: 153–70.

Berndt R. K., Busche G. (1991). *Vogelwelt Schleswig-Holsteins*, Band 3: Entenvögel 1. Karl Wachholtz Verlag, Neumünster, Germany.

Bernstein I. S., Ehardt C. (1985). Agonistic aiding: kinship, rank, age, and sex influences. *American Journal of Primatology*, 8: 37–52.

Bernstein I. S., Ehardt C. (1986). The influence of kinship and socialization on aggressive behaviour in rhesus monkeys (*Macaca mulatta*). *Animal Behaviour*, 34: 739–47.

Bird C. D., Emery N. J. (2009). Insightful problem solving and creative tool modification by captive nontool-using rooks. *Proceedings of the National Academy of Sciences USA*, 106: 10370–5.

Birkhead T. R. (1980). Mate guarding in birds: conflicting interests of males and females. *Animal Behaviour*, 29: 304–5.

Birkhead T. R. (2012). *Bird Sense: What It's Like to Be a Bird*. Bloomsbury, London.

Birkhead T. R., Johnson S. D., Nettleship D. N. (1985). Extra-pair matings and mate guarding in the common murre *Uria aalge*. *Animal Behaviour*, 33: 608–19.

Björklund M., Westman B. (1986). Adaptive advantages of monogamy in great tits (*Parus major*): an experimental test of the polygyny threshold model. *Animal Behaviour*, 34: 1436–40.

Black J. M. (1996). *Partnership in Birds: The Study of Monogamy*. Oxford University Press, Oxford.

Black J. M. (2001). Fitness consequences of long-term pair bonds in barnacle geese: monogamy in the extreme. *Behavioral Ecology*, 12: 640–5.

Black J. M., Owen M. (1987). Determinants of social rank in goose flocks: acquisition of social rank in young geese. *Behaviour*, 102: 129–46.

Black J. M., Owen M. (1989a). Parent–offspring relationships in wintering barnacle geese. *Animal Behaviour*, 37: 187–98.

Black J. M., Owen M. (1989b). Agonistic behaviour in goose flocks: assessment, investment, and reproductive success. *Animal Behaviour*, 37: 199–209.

Black J. M., Owen M. (1995). Reproductive performance and assortative pairing in relation to age in barnacle geese. *Journal of Animal Ecology*, 64: 234–44.

Black J. M., Carbone C., Wells R. L., *et al.* (1992). Foraging dynamics in goose flocks: the cost of living on the edge. *Animal Behaviour*, 44: 41–50.

Black J. M., Choudhury S., Owen M. (1996). Do barnacle geese benefit from lifelong monogamy? In Black J. M. (ed.) *Partnership in Birds: The Study of Monogamy*. Oxford University Press, Oxford, pp. 91–117.

Black J. M., Prop J., Larsson K. (2007). *Wild Goose Dilemmas: Population Consequences of Individual Decisions in Barnacle Geese*. Branta Press, Groningen, Netherlands.

Blix A. S., Stromme S. B., Ursin H. (1974). Additional heart rate: an indicator of psychological activation. *Aerospace Medicine*, 45: 1219–22.

Boag D. A., Alway J. H. (1980). Effect of social environment within the brood on dominance rank in gallinaceous birds (Tetraonidae and Phasanidae). *Canadian Journal of Zoology*, 58: 44–9.

Boissy A., LeNeindre P. (1997). Behavioral, cardiac and cortisol responses to brief peer separation in cattle. *Physiology and Behavior*, 61: 693–9.

Bond A. B., Kamil A. C., Balda R. P. (2003). Social complexity and transitive inference in corvids. *Animal Behaviour*, 65: 479–87.

Bond A. B., Wei C. A., Kamil A. C. (2010). Cognitive representation in transitive inference: a comparison of four corvid species. *Behavioural Processes*, 85: 283–92.

Bonnie K. E., de Waal F. B. M. (2006). Affiliation promotes the transmission of a social custom: handclasp grooming among captive chimpanzees. *Primates*, 47: 27–34.

Bortolotti G. R., Marchant T. A., Blas J., *et al.* (2008). Corticosterone in feathers is a long-term integrated measure of avian stress physiology. *Functional Ecology*, 22: 494–500.

Both C., Dingemanse N. J., Drent P. J., *et al.* (2005). Pairs of extreme avian personalities have highest reproductive success. *Journal of Animal Ecology*, 74: 667–74.

Bowler J., Mitchell C., Leitch A. J. (2005). Greylag geese on Tiree and Coll, Scotland: status, habitat use and movements. *Waterbirds*, 28: 61-70.

Boyd H. (1953). On encounters between wild white-fronted geese in winter flocks. *Behaviour*, 5: 85–130.

Bradley J. S., Wooller R. D., Skira I. J. (1995). The relationships of pair-bond formation and duration to reproductive success in short-tailed shearwaters *Puffinus tenuirostris*. *Animal Ecology*, 64: 31–8.

Bradley M. M., Lang P. J. (2000). Measuring emotion: behavior, feeling, and physiology. In Lane R. D., Nadel L. (eds.) *Cognitive Neuroscience of Emotion*. Oxford University Press, Oxford, pp. 242–76.

Bradley M. M., Codispoti M., Cuthbert B., *et al.* (2001). Emotion and motivation. I. Defensive and appetitive reactions in picture processing. *Emotion*, 1: 276–98.

Brent L. J. N., Semple S., Dubuc C., *et al.* (2011). Social capital and physiological stress levels in free-ranging adult female rhesus macaques. *Physiology and Behavior*, 102: 76–83.

Brown C. R., Brown M. B., Shaffer M. L. (1991). Food-sharing signals among socially foraging cliff swallows. *Animal Behaviour*, 42: 551–64.

Brown J. L. (1975). *The Evolution of Behavior.* W. W. Norton & Company Inc., New York.

Brown J. L. (1987). *Helping and Communal Breeding in Birds.* Princeton University Press, Princeton, NJ.

Bshary R., Wickler W., Fricke H. (2002). Fish cognition: a primate's eye view. *Animal Cognition*, 5: 1–13.

Bugnyar T. (2011). Knower–guesser differentiation in ravens: others' viewpoints matter. *Proceedings of the Royal Society B*, 278: 634–40.

Bugnyar T., Heinrich B. (2005). Ravens, *Corvus corax*, differentiate between knowledgeable and ignorant competitors. *Proceedings of the Royal Society B*, 272: 1641–6.

Bugnyar T., Kotrschal K. (2002). Observational learning and the raiding of food caches in ravens, *Corvus corax*: is it 'tactical' deception? *Animal Behaviour*, 64: 185–95.

Bugnyar T., Kotrschal K. (2004). Leading a conspecific away from food in ravens (*Corvus corax*)? *Animal Cognition*, 7: 69–76.

Bugnyar T., Stöwe M., Heinrich B. (2004). Ravens, *Corvus corax*, follow gaze direction of humans around obstacles. *Proceedings of the Royal Society B*, 271: 1331–6.

Burish M. J., Kueh H. Y., Wang S. S.-H. (2004). Brain architecture and social complexity in modern and ancient birds. *Brain, Behavior and Evolution*, 63: 107–24.

Burton J. A., Risdon D. H. S. (1987). *Vögel in Farbe.* Südwest Verlag, Munich.

Byrne R. W., Whiten A. (1988). *Machiavellian Intelligence: Social Expertise and the Evolution of Intellect in Monkeys, Apes, and Humans.* Clarendon Press, Oxford.

Cabanac A., Cabanac M. (2000). Heart rate responses to gentle handling of frog and lizard. *Behavioural Processes*, 52: 89–95.

Candland D. K., Taylor D. B., Dresdale L., *et al.* (1969). Heart rate, aggression, and dominance in the domestic chicken. *Journal of Comparative and Physiological Psychology*, 67: 70–6.

Candland D. K., Bryan D. C., Nazar B. L., *et al.* (1970). Squirrel monkey heart rate during formation of status orders. *Journal of Comparative and Physiological Psychology*, 70: 417–23.

Carere C., van Oers K. (2004). Shy and bold great tits (*Parus major*): body temperature and breath rate in response to handling stress. *Physiology and Behavior*, 82: 905–12.

Carere C., Groothuis T. G. G., Möstl E., *et al.* (2003). Fecal corticosteroids in a territorial bird selected for different personalities: daily rhythm and the response to social stress. *Hormones and Behavior*, 43: 540–8.

Carere C., Drent P. J., Privitera L., *et al.* (2005). Personalities in great tits, *Parus major*: stability and consistency. *Animal Behaviour*, 70: 795–805.

Caro S. P., Charmantier A., Lambrechts M. M., *et al.* (2009). Local adaptation of timing of reproduction: females are in the driver's seat. *Functional Ecology*, 23: 172–9.

Carter C. S. (1992). Neuroendocrinology of sexual behaviour in the female. In Becker J. B., Breedlove S. M., Crews D. (eds.) *Behavioral Endocrinology.* Bradford Books, MIT Press, Cambridge, MA, pp. 71–96.

Cezilly F., Nager R. G. (1996). Age and breeding performance in monogamous birds: the influence of pair stability. *Trends in Ecology and Evolution*, 11: 27.

Cheney D. L., Seyfarth R. M. (1981). Selective forces affecting the predator alarm calls of vervet monkeys. *Behaviour*, 76: 25–61.

Cheney D. L., Seyfarth R. M. (1986). The recognition of social alliances by vervet monkeys. *Animal Behaviour*, 34: 1722–31.

Cheney D. L., Seyfarth R. M. (2003). The structure of social knowledge in monkeys. *In* de Waal F. B. M., Tyack P. L. (eds.) *Animal Social Complexity.* Harvard University Press, Cambridge, MA, pp. 207–29.

Cheney D. L., Seyfarth R. M., Smuts B. (1986). Social relationships and social cognition in nonhuman primates. *Science*, 234: 1361–6.

Choudhury S., Black J. M. (1993). Mate-selection behaviour and sampling strategies in geese. *Animal Behaviour*, 46: 747–57.

Choudhury S., Jones C. S., Black J. M., *et al.* (1993). Adoption of young and intraspecific nest parasitism in barnacle geese. *Condor*, 95: 860–8.

Choudhury S., Black J. M., Owen M. (1996). Body size, fitness and compatibility in barnacle geese *Branta leucopsis*. *Ibis*, 138: 700–9.

Christenfeld N., Gerin W., Linden W., *et al.* (1997). Social support effects on cardiovascular reactivity: is a stranger as effective as a friend? *Psychosomatic Medicine*, 59: 388–98.

Clark C. W., Mangel M. (1986). The evolutionary advantages of group foraging. *Theoretical Population Biology*, 30: 45–75.

Clarke J. A., Tambussi C. P., Noriega J. I., *et al.* (2005). Definitive fossil evidence for the extant avian radiation in the Cretaceous. *Nature*, 433: 305–8.

Clayton N. S., Dickinson A. (1998). Episodic-like memory during cache recovery by scrub jays. *Nature*, 395: 272–4.

Clifton P. G., Andrew R. J., Brighton L. (1988). Gonadal steroids and attentional mechanism in young domestic chicks. *Physiology and Behavior*, 43: 441–6.

Clutton-Brock T., Sheldon B. C. (2010). The seven ages of *Pan*. *Science*, 327: 1207–8.

Cockburn A. (1996). Why do so many Australian birds cooperate? Social evolution in the Corvida. *In* Floyd R., Sheppard A., de Barro P. (eds.) *The Nicholson Centenary Meeting: Frontiers of Population Ecology*, Melbourne, Australia.

Cockrem J. F., Barrett D. P., Candy E. J., *et al.* (2009). Corticosterone responses in birds: individual variation and repeatability in Adelie penguins (*Pygoscelis adeliae*) and other species, and the use of power analysis to determine sample sizes. *General and Comparative Endocrinology*, 163: 158–68.

Coleman K., Wilson D. S. (1998). Shyness and boldness in pumpkinseed sunfish: individual differences are context-specific. *Animal Behaviour*, 56: 927–36.

Conradt L., Roper T. J. (2005). Consensus decision making in animals *Trends in Ecology and Evolution*, 20: 449–56.

Cooke F., Rockwell R. F., Lank D. B. (1995). *The Snow Geese of La Perouse Bay*. Oxford University Press, Oxford.

Cords M., Aureli F. (2000). Reconciliation and relationship qualities. *In* Aureli F., de Waal F. B. M. (eds.) *Natural Conflict Resolution*. Berkeley University Press, Berkeley, CA.

Côté I., Poulinb R. (1995). Parasitism and group size in social animals: a meta-analysis. *Behavioral Ecology*, 6: 159–65.

Creel S. (2001). Social dominance and stress hormones. *Trends in Ecology and Evolution*, 16: 491–7.

Creel S., Spong G., Sands J. L., *et al.* (2003). Population size estimation in Yellowstone wolves with error-prone non-invasive microsatellite genotypes. *Molecular Ecology*, 12: 2003–9.

Croft D. P., Krause J., Richard J. (2004). Social networks in the guppy (*Poecilia reticulata*). *Biology Letters*, 271: S516–S519.

Croft D. P., James R., Botham A. J. W., *et al.* (2005). Assortative interactions and social networks in fish. *Oecologia*, 143: 211–19.

Croft D. P., James R., Thomas P. O. R., *et al.* (2006). Social structure and co-operative interactions in a wild population of guppies (*Poecilia reticulata*). *Behavioral Ecology and Sociobiology*, 59: 644–50.

Curley J. P., Keverne E. B. (2005). Genes, brains and mammalian social bonds. *Trends in Ecology and Evolution*, 20: 561–7.

Daisley J. N., Kotrschal K. (2000). Testosterone and development of behavioural phenotypes in precocial birds. *In* 7th International Symposium on Avian Endocrinology, Varanasi, India.

Daisley J. N., Bromundt V., Möstl E., *et al.* (2005). Enhanced yolk testosterone influences behavioral phenotype independent of sex in Japanese quail chicks *Coturnix japonica*. *Hormones and Behavior*, 47: 185–94.

Dally J., Emery N. J., Clayton N. S. (2006). Food-caching Western scrub jays keep track of who was watching when. *Science*, 312: 1662–5.

Darwin C. (1859). *On the Origin of Species by Means of Natural Selection, or the Preservation of Favoured Races in the Struggle for Life*. John Murray, London.

Darwin C. (1872). *The Expression of the Emotions in Man and Animals*. John Murray, London.

Datta S. B. (1983a). Relative power and the acquisition of rank. *In* Hinde R. A. (ed.) *Primate Social Relationships*. Blackwell Science, Boston, MA, pp. 93–102.

Datta S. B. (1983b). Relative power and the maintenance of dominance. *In* Hinde R. A. (ed.) *Primate Social Relationships*. Blackwell Science, Boston, MA, pp. 103–11.

Davis E. S. (2002). Female choice and the benefits of mate guarding by male mallards. *Animal Behaviour*, 64: 619–28.

Davis H. (1992). Transitive inference in rats (*Rattus norvegicus*). *Journal of Comparative Psychology*, 106: 342–9.

Dawkins R. (1980). Good strategy or evolutionary stable strategy? *In* Barlow G. W., Silverberg J. (eds.) *Sociobiology: Beyond Nature/Nurture?* Westview Press, Boulder, CO.

de Kort S. M., Emery N. J., Clayton N. S. (2003). Food offering in jackdaws, *Corvus monedula*. *Naturwissenschaften*, 90: 238–40.

de Kort S. M., Emery N. J., Clayton N. S. (2006). Food sharing in jackdaws, *Corvus monedula*: what, why and with whom? *Animal Behaviour*, 72: 297–304.

Delius J. D., Siemann M., Emmerton J., *et al.* (2001). Cognition of birds as products of evolved brains. *In* Roth G., Wullimann M. F. (eds.) *Brain Evolution and Cognition*. Wiley & Sons, New York, pp. 451–90.

De Palma C., Viggiano E., Barillari E., *et al.* (2005). Evaluating the temperament in shelter dogs. *Behaviour*, 142: 1307–28.

Desire L., Veissier I., Despres G., *et al.* (2004). On the way to assess emotions in animals: do lambs (*Ovis aries*) evaluate an event through its suddenness, novelty, or unpredictablity? *Journal of Comparative Psychology*, 118: 363–74.

Desire L., Veissier I., Despres G., *et al.* (2006). Appraisal process in sheep (*Ovis aries*): interactive effect of suddenness and unfamiliarity on cardiac and behavioral responses. *Journal of Comparative Psychology*, 120: 280–7.

DeVries A. C. (2002). Interaction among social environment, the hypothalamic-pituitary-adrenal axis, and behavior. *Hormones and Behavior*, 41: 405–13.

DeVries A. C., Glasper E. F., Detillion C. E. (2003). Social modulation of stress responses. *Physiology and Behavior*, 79: 399–407.

Dingemanse N. J., Réale D. (2005). Natural selection and animal personality. *Behaviour*, 142: 1159–84.

Dingemanse N. J., Van der Plas F., Wright J., *et al.* (2009). Individual experience and evolutionary history of predation affect expression of heritable variation in fish personality and morphology. *Proceedings of the Royal Society B*, 276: 1285–93.

Dittami J. P. (1981). Seasonal changes in the behavior and plasma titers of various hormones in barheaded geese, *Anser indicus*. *Zeitschrift für Tierpsychologie*, 55: 289–324.

Dixon A., Ross D. J., O'Malley S. L. C., *et al.* (1994). Paternal investment inversely related to degree of extra-pair paternity in the reed bunting. *Nature*, 371: 698–700.

Dobberfuhl A. P., Ullmann J. F. P., Shumway C. A. (2005). Visual acuity, environmental complexity, and social organization in African cichlid fishes. *Behavioral Neuroscience*, 119: 1648–55.

Drea C. M., Frank L. G. (2003). The social complexity of spotted hyenas. In de Waal F. B. M., Tyack P. L. (eds.) *Animal Social Complexity: Intelligence, Culture, and Individualized Societies*. Harvard University Press, Cambridge, MA, pp. 121–48.

Drent P. J., Marchetti C. (1999) Individuality, exploration and foraging in hand raised juvenile great tits. In Adams N. J., Slotow R. H. (eds.) *22nd International Ornithological Congress*, Durban, South Africa, S16.4.

Dressen W., Grün H., Hendrichs H. (1990). Radio telemetry of heart rate in male tammar wallabies (Marsupialia: Macropodidae): temporal variations and behavioural correlates. *Australian Journal of Zoology*, 38: 89–103.

Dugatkin L. A. (1997). *Cooperation Among Animals*. Oxford University Press, Oxford.

Dukas R., Kamil A. C. (2000). The cost of limited attention in blue jays. *Behavioral Ecology*, 11: 502–6.

Dunbar R. I. M. (1980). Determinants and evolutionary consequences of dominance among Gelada baboons. *Behavioral Ecology and Sociobiology*, 7: 263–5.

Dunbar R. I. M. (1992). Neocortex size as a constraint on group size in primates. *Journal of Human Evolution*, 22: 469–93.

Dunbar R. I. M. (1995). Neocortex size and group size in primates: a test of the hypothesis. *Journal of Human Evolution*, 28: 287–96.

Dunbar R. I. M. (1998). The social brain hypothesis. *Evolutionary Anthropology*, 6: 178–90.

Dunbar R. I. M., Bever J. (1998). Neocortex size predicts group size in carnivores and some insectivores. *Ethology*, 104: 695–708.

Dunbar R. I. M., Shultz S. (2007). Evolution in the social brain. *Science*, 317: 1344–7.

Dunn P. O., Afton A. D., Gloutney M. L., *et al.* (1999). Forced copulation results in few extrapair fertilizations in Ross's and Lesser Snow Geese. *Animal Behaviour*, 57: 1071–81.

Dzus E. H., Clark R. G. (1996). Effects of harness-style and abdominally implanted transmitters on survival and return rates of mallards. *Journal of Field Ornithology*, 67: 549–57.

Eadie J. M., Kehoe F. P., Nudds T. D. (1988). Pre-hatch and post-hatch brood amalgamation in North American Anatidae: a review of hypotheses. *Canadian Journal of Zoology*, 66: 1709–21.

Ebinger P., Löhmer R. (1987). A volumetric comparison of brains between greylag geese (*Anser anser* L.) and domestic geese. *Journal of Brain Research*, 28: 291–9.

Edinger T. (1929). Die fossilen Gehirne. *Zeitschrift für die Gesamte Anatomie*, Abt. 3., 28: 1–249.

Eisermann K. (1992). Long-term heart-rate responses to social stress in wild European rabbits: predominant effect of rank position. *Physiology and Behavior*, 52: 33–6.

Eising C. M., Eikenaar C., Schwabl H., *et al.* (2001). Maternal androgens in black-headed gull (*Larus ridibundus*) eggs: consequences for chick development. *Proceedings of the Royal Society B*, 268: 839–46.

Ekman J., Bylin A., Tegelström H. (2000). Parental nepotism enhances survival of retained offspring in the Siberian jay. *Behavioral Ecology*, 11: 416–20.

Elder W. H., Elder N. L. (1949). Role of the family in the formation of goose flocks. *Wilson Bulletin*, 61: 133–40.

Ely C. R. (1993). Family stability in greater white-fronted geese. *Auk*, 110: 425–35.

Ely C. R., Ward D. H., Bollinger K. S. (1999). Behavioral correlates of heart rates of free-living greater white-fronted geese. *Condor*, 101: 390–5.

Emery N. J. (2000). The eyes have it: the neuroethology, function and evolution of social gaze. *Neuroscience and Biobehavioral Reviews*, 24: 581–604.

Emery N. J., Clayton N. S. (2004). The mentality of crows: convergent evolution of intelligence in corvids and apes. *Science*, 306: 1903–7.

Emery N. J., Seed A. M., von Bayern A. M. P., *et al.* (2007). Cognitive adaptations of social bonding in birds. *Philosophical Transactions of the Royal Society B*, 362: 489–505.

Emlen S. T. (1997). Family dynamics in social vertebrates. *In* Krebs J. R., Davies N. B. (eds.) *Behavioural Ecology: An Evolutionary Approach*, 4th edn. Wiley-Blackwell, Malden, MA, pp. 228–53.

Emlen S. T., Oring L. W. (1977). Ecology, sexual selection, and the evolution of mating systems. *Science*, 197: 215–23.

Engh A. L., Siebert E. R., Greenberg D. A., *et al.* (2005). Patterns of alliance formation and postconflict aggression indicate spotted hyaenas recognize third party relationships. *Animal Behaviour*, 69: 209–17.

Estevez I., Newberry R. C., Arias de Reyna L. (1997). Broiler chickens: a tolerant social system? *Etologia*, 5: 19–29.

Estevez I., Keeling L. J., Newberry R. C. (2003). Decreasing aggression with increasing group size in young domestic fowl. *Applied Animal Behaviour Science*, 84: 213–18.

Fairbanks L. A., Newman T. K., Bailey J. N., *et al.* (2004). Genetic contributions to social impulsivity and aggressiveness in vervet monkeys. *Biological Psychiatry*, 55: 642–7.

Falconer D. S. (1952). The problem of environment and selection. *American Naturalist*, 86: 293–8.

Fallani G., Prato Previde E., Valsecchi P. (2007). Behavioral and physiological responses of guide dogs to a situation of emotional distress. *Physiology and Behavior*, 90: 648–55.

Farago S. (2010). Numbers and distributions of geese in Hungary 1984–2009. *Ornis Svecica*, 20: 144–54.

Fedigan L. M. (1992). *Primate Paradigms: Sex Roles and Social Bonds*. University of Chicago Press, Chicago, IL.

Feige N., van der Jeugd H. P., Voslamber B., *et al.* (2008). Characterisation of Greylag Goose *Anser anser* breeding areas in the Netherlands with special regard to human land use. *Vogelwelt*, 129: 348–59.

Fels D., Ap Rhisiart A., Vollrath F. (1995). The selfish crouton. *Behaviour*, 132: 49–56.

Ficken M. S., Weise C. M., Popp J. W. (1990). Dominance rank and resource access in winter flocks of black-capped chickadees. *Wilson Bulletin*, 102: 623–33.

Filby A. L., Paull G. C., Bartlett E. J., et al. (2010). Physiological and health consequences of social status in zebrafish (Danio rerio). Physiology and Behavior, 101: 576–87.

Fischer H. (1965). Das Triumphgeschrei der Graugans (Anser anser). Zeitschrift für Tierpsychologie, 22: 247–304.

Fischer H. (1973a) Soziales Verhalten der Graugans: Paarbildung. In Wissenschaftlicher Film D 975/1967. Institut für den Wissenschaftlichen Film, Göttingen, Germany.

Fischer H. (1973b) Soziales Verhalten der Graugans: Fortpflanzung. In Wissenschaftlicher Film D 976/1967. Institut für den Wissenschaftlichen Film, Göttingen, Germany.

Fisher R. A. (1958). The Genetical Theory of Natural Selection, 2nd edn. Dover, New York.

Föger B., Taschwer K. (2001). Die andere Seite des Spiegels. Konrad Lorenz und der Nationalsozialismus. Czernin Verlag, Vienna.

Fokkema D. S., Koolhaas J. M. (1985). Acute and conditioned blood pressure changes in relation to social and psychosocial defeat in rats. Physiology and Behavior, 34: 33–8.

Fowler G. S. (1995). Stages of age related reproductive success in birds: simultaneous effects of age, pair-bond duration and reproductive experience. American Zoologist, 35: 318–28.

Fox A. D., Boyd H., Bromley R. G. (1995). Mutual benefits of associations between breeding and non-breeding White-fronted Geese, Anser albifrons. Ibis, 137: 151–6.

Fox A. D., Glahder C. M., Walsh A. J. (2003). Spring migration routes and timing of Greenland white-fronted geese: results from satellite telemetry. OIKOS, 103: 415–25.

Fox A. D., Ebbinge B. S., Mitchell C., et al. (2010). Current estimates of goose population sizes in western Europe, a gap analysis and an assessment of trends. Ornis Svecica, 20: 115–27.

Frank M. J., Rudy J. W., Levy W. B., et al. (2005). When logic fails: implicit transitive inference in humans. Memory and Cognition, 33: 742–50.

Fraser G. S., Jones I. L., Hunter F. M. (2002). Male–female differences in parental care in monogamous crested auklets. Condor, 104: 413–23.

Fraser O. N., Bugnyar T. (2010a). The quality of social relationships in ravens. Animal Behaviour, 79: 927–33.

Fraser O. N., Bugnyar T. (2010b). Do ravens show consolation? Responses to distressed others. PLoS One, 5: e10605.

Frederiksen M., Hearn R. D., Mitchell C., et al. (2004). The dynamics of hunted Icelandic goose populations: a reassessment of the evidence. Journal of Applied Ecology, 41: 315–34.

Fricke H. (1975a). Sozialstruktur und ökologische Spezialisierung von verwandten Fischen (Pomacentridae). Zeitschrift für Tierpsychologie, 39: 492–520.

Fricke H. (1975b). Lösen einfacher Probleme bei einem Fisch. Zeitschrift für Tierpsychologie, 38: 18–33.

Frigerio D., Weiß B. M., Kotrschal K. (2001a). Spatial proximity among adult siblings in greylag geese (Anser anser): evidence for female bonding? Acta Ethologica, 3: 121–5.

Frigerio D., Möstl E., Kotrschal K. (2001b). Excreted metabolites of gonadal steroid hormones and corticosterone in greylag geese (Anser anser) from hatching to fledging. General and Comparative Endocrinology, 124: 246–55.

Frigerio D., Weiß B. M., Dittami J., et al. (2003). Social allies modulate corticosterone excretion and increase success in agonistic interactions in juvenile hand-raised greylag geese (Anser anser). Canadian Journal of Zoology, 81: 1746–54.

Frigerio D., Hirschenhauser K., Möstl E., *et al.* (2004a). Experimentally elevated testosterone increases status signalling in male Greylag geese (*Anser anser*). *Acta Ethologica*, 7: 9–18.

Frigerio D., Dittami J., Möstl E., *et al.* (2004b). Excreted corticosterone metabolites co-vary with ambient temperature and air pressure in male Greylag geese (*Anser anser*). *General and Comparative Endocrinology*, 137: 29–36.

Fritz J., Kotrschal K. (2002). On avian imitation: cognitive and ethological perspectives. *In* Dauterhan K., Nehaniv C. L. (eds.) *Imitation in Animals and Artefacts*. MIT Press, Cambridge, MA, pp. 133–56.

Fritz J., Bisenberger A., Kotrschal K. (1999). Social mediated learning of an operant task in greylag geese: field observation and experimental evidence. *Advances in Ethology*, 34: 51.

Fritz J., Bisenberger A., Kotrschal K. (2000). Stimulus enhancement in greylag geese: socially mediated learning of an operant task. *Animal Behaviour*, 59: 1119–25.

Fucikova E., Drent P. J., Smits N., *et al.* (2009). Handling stress as a measurement of personality in great tit nestlings (*Parus major*). *Ethology*, 115: 366–74.

Fuxe K., Diaz R., Cintra A., *et al.* (1996). On the role of glucocorticoid receptors in brain plasticity. *Cellular and Molecular Neurobiology*, 16: 239–58.

Galsworthy M. J., Dionne G., Dale P. S., *et al.* (2000). Sex differences in early verbal and non-verbal cognitive development. *Developmental Science*, 3: 206–15.

Garamszegi L. Z., Calhim S., Dochtermann N. A., *et al.* (2009). Changing philosophies and tools for statistical inferences in behavioral ecology. *Behavioral Ecology*, 20: 1363–75.

Gauthier G., Tardif J. (1991). Female feeding and male vigilance during nesting in greater snow geese. *Condor*, 93: 701–11.

Gauthier G., Pradel R., Menu S., *et al.* (2001). Seasonal survival of Greater Snow Geese and effect of hunting under dependence in sighting probability. *Ecology*, 82: 3105–19.

Geffen E., Yom-Tov Y. (2001). Factors affecting the rates of intraspecific nest parasitism among Anseriformes and Galliformes. *Animal Behaviour*, 62: 1027–38.

Gehrt S. D., Fritzell E. K. (1998). Duration of familial bonds and dispersal patterns for raccoons in south Texas. *Journal of Mammalogy*, 79: 859–72.

Gil D., Graves J., Wells A. (1999). Male attractiveness and differential testosterone investment in zebra finch eggs. *Science*, 286: 126–8.

Gill F. B. (1994). *Ornithology*, 2nd edn. W. H. Freeman & Co., New York.

Gillan D. J. (1981). Reasoning in chimpanzees. II. Transitive inference. *Journal of Experimental Psychology: Animal Behavior Processes*, 7: 150–64.

Giraldeau L.-A. (1997). The ecology of information use. *In* Krebs J. R., Davies N. B. (eds.) *Behavioral Ecology*, 4th edn. Blackwell Science, Oxford, pp. 42–68.

Gittleman J. L. (1986). Carnivore brain size, behavioral ecology and phylogeny. *Journal of Mammalogy*, 67: 23–36.

Goodall J. (1986). *The Chimpanzees of Gombe: Patterns of Behavior*. Belknap Press, Harvard University Press, Cambridge, MA.

Goodson J. L. (2005). The vertebrate social behavior network: evolutionary themes and variations. *Hormones and Behavior*, 48: 11–22.

Gosling S. D. (2001). From mice to men: what can we learn about personality from animal research? *Psychological Bulletin*, 127: 45–86.

Gosling S. D., John O. P. (1999). Personality dimensions in nonhuman animals: a cross-species review. *Current Directions in Psychological Science*, 8: 69–75.

Goymann W., Wingfield J. C. (2004). Allostatic load, social status and stress hormones: the costs of social status matter. *Animal Behaviour*, 67: 591–602.

Grammer K., Kruck K. B., Magnusson M. S. (1998). The courtship dance: patterns of nonverbal synchronization in opposite-sex encounters. *Journal of Nonverbal Behavior*, 22: 3–29.

Gregoire P. E., Ankney C. D. (1990). Agonistic behavior and dominance relationships among lesser snow geese during winter and spring migration. *Auk*, 107: 550–60.

Griffith S. C., Owens I. P. F., Thuman K. A. (2002). Extra pair paternity in birds: a review of intraspecific variation and adaptive function. *Molecular Ecology*, 11: 2195–212.

Griffith S. C., Pryke S. R., Buttemer W. A. (2011). Constrained mate choice in social monogamy and the stress of having an unattractive partner. *Proceedings of the Royal Society B*, 278: 2798–805.

Groothuis T. G. G., Carere C. (2005). Avian personalities: characterization and epigenesis. *Neuroscience*, 29: 137–50.

Grosenick L., Clement T. S., Fernald R. D. (2007). Fish can infer rank by observation alone. *Nature*, 445: 429–32.

Grubb B. R. (1983). Allometric relations of cardiovascular function in birds. *American Journal of Physiology: Heart and Circulatory Physiology*, 245: H567–H572.

Gruber-Baldini A. L., Schaie K. W., Willis S. L. (1995). Similarity in married couples: a longitudinal study of mental abilities and rigidity/flexibility. *Journal of Personality and Social Psychology* 69: 191–203.

Gubler H. (1989). *Paarbindung und Distanzregulation bei der Chinesischen Zwergwachtel* (*Excalfactoria chinensis*). Dissertation, University of Zurich.

Güntürkün O. (2005). The avian 'prefrontal cortex' and cognition. *Current Opinion in Neurobiology*, 15: 686–93.

Gvaryahu G., Snapir N., Robinzon B., *et al.* (1986). The gonadotrophic-axis involvement in the course of the filial following response in the domestic fowl chick. *Physiology and Behavior*, 38: 651–6.

Gwinner E. (1964). Untersuchungen über das Ausdrucks- und Sozialverhalten des Kolkraben (*Corvus corax corax* L.). *Zeitung für Tierpsychologie*, 21: 657–748.

Gygax L., Neisen G., Wechsler B. (2010). Socio-spatial relationships in dairy cows. *Ethology*, 116: 10–23.

Hagelin J. C., Jones I. L. (2007). Bird odors and other chemical substances: a defense mechanism or overlooked mode of intraspecific communication? *Auk*, 124: 741–61.

Hamilton W. D. (1964a). The genetical evolution of social behaviour. I. *Journal of Theoretical Biology*, 7: 1–16.

Hamilton W. D. (1964b). The genetical evolution of social behaviour. II. *Journal of Theoretical Biology*, 7: 17–52.

Hamilton W. D. (1971). Geometry of the selfish herd. *Journal of Theoretical Biology*, 31: 295–311.

Hamre B. K., Pianta R. C. (2005). Can instructional and emotional support in the first-grade classroom make a difference for children at risk of school failure? *Child Development*, 76: 949–67.

Hanson H. C. (1953). Inter-family dominance in Canada geese. *Auk*, 70: 11–16.

Hare B., Agnetta B., Tomasello M. (2000). Chimpanzees know what conspecifics do and do not see. *Animal Behaviour*, 59: 771–86.

Hare B., Call J., Tomasello M. (2001). Do chimpanzees know what conspecifics know? *Animal Behaviour*, 61: 139–51.

Healy S. D., Rowe C. (2007). A critique of comparative studies of brain size. *Proceedings of the Royal Society B*, 274: 453–64.

Hebblewhite M., Haydon D. T. (2010). Distinguishing technology from biology: a critical review of the use of GPS telemetry data in ecology. *Philosophical Transactions of the Royal Society B*, 365: 2303–12.

Heg D., van Treuren R. (1998). Female-female cooperation in polygynous oystercatchers. *Nature*, 391: 687–91.

Heg D., Bruinzeel L. W., Ens B. J. (2003). Fitness consequences of divorce in the oystercatcher, *Haematopus ostralegus*. *Animal Behaviour*, 66: 175–84.

Hegner R. E., Wingfield J. C. (1987). Effects of experimental manipulation of testosterone levels on parental investment and breeding success in male house sparrows. *Auk*, 104: 462–9.

Heinrich B. (1989). *Ravens in Winter*. Summit Books, New York.

Heinrich B. (1999). *Mind of the Raven*. Harper, New York.

Heinrich B., Pepper J. W. (1998). Influence of competitors on caching behaviour in the common raven, *Corvus corax*. *Animal Behaviour*, 56: 1083–90.

Heinrichs M., Baumgartner T., Kirschbaum C., *et al.* (2003). Social support and oxytocin interact to suppress cortisol and subjective responses to psychosocial stress. *Biological Psychiatry*, 54: 1389–98.

Heinroth O. (1911). *Beiträge zur Biologie, namentlich Ethologie and Psychologie der Anatiden*. I: Verhandlungen V. Internationaler Ornithologischer Kongress, Berlin 1910, pp. 589–702.

Hemetsberger J. (2001). Die Entwicklung der Grünauer Graugansschar seit 1973. *In* Kotrschal K., Müller G., Winkler H. (eds.) *Konrad Lorenz und seine verhaltensbiologischen Konzepte aus heutiger Sicht*. Filander Verlag, Fürth, Germany, pp. 249–60.

Hemetsberger J. (2002). *Populationsbiologische Aspekte der Grünauer Graugansschar (Anser anser)*. Dissertation, Institute of Zoology, University of Vienna.

Hemetsberger J., Scheiber I. B. R., Weiß B. M., *et al.* (2010). Influence of socially involved hand-raising on life history and stress responses in greylag geese. *Interaction Studies*, 11: 380–95.

Hennessy M. B., Ritchey R. L. (1987). Hormonal and behavioral attachment responses in infant guinea pigs. *Developmental Psychobiology*, 20: 613–25.

Hess E. (1973). *Imprinting: Early Experience and the Developmental Psychobiology of Attachment*. D. Van Nostrand Company, New York.

Hessing M. J. C., Hagelso A. M. (1993). Individual behavioral characteristics in pigs. *Applied Animal Behaviour Science*, 37: 285–95.

Hessing M. J. C., Scheepends C. J. M., Schouten W. G. P., *et al.* (1994). Social rank and disease susceptibility in pigs. *Veterinary Immunology and Immunopathology*, 43: 373–87.

Hill D. A. (1997). Seasonal variation in the feeding behavior and diet of Japanese macaques (*Macaca fuscata yakui*) in lowland forest of Yakushima. *American Journal of Primatology*, 43: 305–22.

Hinde R. A. (1979). *Towards Understanding Relationships*. Academic Press, London.

Hinde R. A. (1983). *Primate Social Relationships: An Integrated Approach*. Blackwell Press, Oxford.

Hirschenhauser K. (1998). *Steroidhormone aus Kot und Sozialverhalten bei Graugänsen*. Dissertation, Institute of Zoology, University of Vienna.

Hirschenhauser K. (2012). Testosterone and partner compatibility: evidence and emerging questions. *Ethology*, 118: 799–811.

Hirschenhauser K., Oliveira R. F. (2006). Social modulation of androgens in male vertebrates: meta-analyses of the challenge hypothesis. *Animal Behaviour*, 71: 265–77.

Hirschenhauser K., Möstl E., Kotrschal K. (1999a). Seasonal patterns of sex steroids determined from feces in different social categories of Greylag geese (*Anser anser*). *General and Comparative Endocrinology*, 114: 67–79.

Hirschenhauser K., Möstl E., Kotrschal K. (1999b). Within-pair testosterone covariation and reproductive output in greylag geese (*Anser anser*). *Ibis*, 141: 577–86.

Hirschenhauser K., Möstl E., Péczely P., et al. (2000). Seasonal relationships between plasma and fecal testosterone in response to GnRH in domestic ganders. *General and Comparative Endocrinology*, 118: 262–72.

Hirschenhauser K., Taborsky M., Oliveira T., et al. (2004). A test of the 'challenge hypothesis' in cichlid fish: simulated partner and territory intruder experiments. *Animal Behaviour*, 68: 741–50.

Hirschenhauser K., Kotrschal K., Möstl E. (2005). Synthesis of measuring steroid metabolites in goose feces. *Annals of the New York Academy of Sciences*, 1046: 138–53.

Hirschenhauser K., Weiß B. M., Haberl W., et al. (2010). Female androgen patterns and within-pair testosterone compatibility in domestic geese (*Anser anser*). *General and Comparative Endocrinology*, 165: 195–203.

Hogstadt O. (1987). It is expensive to be dominant. *Auk*, 104: 333–6.

Hohnstein A. (2010). *Individualerkennung bei Graugänsen*. Diploma Thesis, Institute of Biology, Humboldt University of Berlin.

Höjesjö J., Johnsson J. I., Petersson E., et al. (1998). The importance of being familiar: individual recognition and social behavior in sea trout (*Salmo trutta*). *Behavioral Ecology*, 9: 445–51.

Holekamp K. E. (2007). Questioning the social intelligence hypothesis. *Trends in Cognitive Sciences*, 11: 65–9.

Horrocks J., Hunte W. (1983). Maternal rank and offspring rank in vervet monkeys: an appraisal of the mechanisms of rank acquisition. *Animal Behaviour*, 31: 772–82.

Huber R., Martys M. (1993). Male–male pairs in greylag geese (*Anser anser*). *Journal of Ornithology*, 134: 155–64.

Hughes W. O. H., Eilenberg J., Boomsma J. J. (2002). Trade-offs in group living: transmission and disease resistance in leaf-cutting ants. *Proceedings of the Royal Society B*, 269: 1811–19.

Humphrey N. K. (1976). The social function of intellect. In Bateson P., Hinde R. (eds.) *Growing Points in Ethology*. Cambridge University Press, Cambridge, pp. 303–21.

Hunt G. R. (1996). Manufacture and use of hook-tools by New Caledonian crows. *Nature*, 379: 249–51.

Huxley J. S. (1914). The courtship habits of the great creasted grebe (*Podiceps cristatus*) with an addition to the theory of sexual selection. *Proceedings of the Royal Society B*, 84: 491–562.

Isler K., van Schaik C. P. (2009). Why are there so few smart mammals (but so many smart birds)? *Biology Letters*, 5: 125–9.

Ito H., Yamamoto N. (2009). Non-laminar cerebral cortex in teleost fishes? *Biology Letters*, 5: 117–21.

Ito H., Ishikawa Y., Yoshimoto M., et al. (2007). Diversity of brain morphology in teleosts: brain and ecological niche. *Brain, Behavior and Evolution*, 69: 76–86.

Iwaniuk A. N., Arnold K. E. (2004). Is cooperative breeding associated with bigger brains? A comparative test in the Corvida (Passeriformes). *Ethology*, 110: 203–20.

Iwaniuk A. N., Hurd P. L. (2005). The evolution of cerebrotypes in birds. *Brain, Behavior and Evolution*, 65: 215–30.

Iwaniuk A. N., Nelson J. E. (2003). Developmental differences are correlated with relative brain size in birds: a comparative analysis. *Canadian Journal of Zoology*, 81: 1913–28.

Iwaniuk A. N., Dean K. M., Nelson J. E. (2005). Interspecific allometry of the brain and brain regions in parrots (Psittaciformes): comparisons with other birds and primates. *Brain, Behavior and Evolution*, 65: 40–59.

Jaatinen K., Jaari S., O'Hara R. B., *et al.* (2009). Relatedness and spatial proximity as determinants of host–parasite interactions in the brood parasitic Barrow's goldeneye (*Bucephala islandica*). *Molecular Ecology*, 18: 2713–21.

Jaatinen K., Öst M., Gienapp P., *et al.* (2011). Differential responses to related hosts by nesting and non-nesting parasites in a brood-parasitic duck. *Molecular Ecology*, 20: 5328–36.

Jacob G., Debrunner R., Gugerli F., *et al.* (2010). Field surveys of capercaillie (*Tetrao urogallus*) in the Swiss Alps underestimated local abundance of the species as revealed by genetic analyses of non-invasive samples. *Conservation Genetics*, 11: 33–44.

Jacobs L. F. (2006). From movement to transitivity: the role of hippocampal parallel maps in configural learning. *Reviews of Neurosciences*, 17: 99–109.

Jamieson I. G., Craig J. L. (1987). Male–male and female–female courtship and copulation behaviour in a communally breeding bird. *Animal Behaviour*, 35: 1251–3.

Jarvis E. D., Güntürkün O., Bruce L., *et al.* (2005). Avian brains and a new understanding of vertebrate brain evolution. *Nature Reviews*, 6: 151–9.

Jensen J., Sæther B.-E., Ringsby T. H., *et al.* (2003). Sexual variation in heritability and genetic correlations of morphological traits in house sparrows (*Passer domesticus*). *Journal of Evolutionary Biology*, 16: 1296–307.

Jensen P. (1995). Individual variation in the behaviour of pigs: noise or functional coping strategies. *Applied Animal Behaviour Science*, 44: 245–55.

John O. P., Robins R. W., Pervin L. A. (2008). *Handbook of Personality: Theory and Research*, 3rd edn. Guilford Press, New York.

Johnson A. L. (2002). Reproduction in the female. *In* Whittow G. C. (ed.) *Sturkie's Avian Physiology*, 5th edn. Academic Press, San Diego, CA, pp. 569–91.

Johnson D. H., Nichols J. D., Schwartz M. D. (1992). Population dynamics of breeding waterfowl. *In* Batt B. D. J., Afton A. D., Anderson M. G., *et al.* (eds.) *Ecology and Management of Breeding Waterfowl*. University of Minnesota Press, Minneapolis, MN, pp. 446–87.

Johnson J., Raveling D. G. (1988). Weak family associations in cackling geese during winter: effects of body size and food resources on goose social organization. *In* Weller M. W. (ed.) *Waterfowl in Winter*. University of Minnesota Press, Minneapolis, MN.

Johnson J. C., Sih A. (2007). Fear, food, sex and parental care: a syndrome of boldness in the fishing spider, *Dolomedes triton*. *Animal Behaviour*, 74: 1131–8.

Jolly A. (1966). Lemur social behavior and primate intelligence. *Science*, 153: 501–7.

Judge P. G., Mullen S. H. (2005). Quadratic postconflict affiliation among bystanders in a hamadryas baboon group. *Animal Behaviour*, 69: 1345–55.

Kagan J., Reznick J., Snidman N. (1988). Biological bases for childhood shyness. *Science*, 240: 167–71.

Kaiser S., Kirtzeck M., Hornschuh G., *et al.* (2003). Sex-specific difference in social support: a study in female guinea pigs. *Physiology and Behavior*, 79: 297–303.

Kalas S. (1977). Ontogenie und Funktion der Rangordnung innerhalb einer Geschwisterschar von Graugänsen (*Anser anser* L.). *Zeitschrift für Tierpsychologie*, 45: 174–98.

Kalchreuter H. (2000). *Das Wasserwild. Verbreitung und Lebensweise: Jagdliche Nutzung und Erhaltung*. Kosmos Verlag, Stuttgart, Germany.

Kalmbach E. (2006). Why do goose parents adopt unrelated goslings? A review of hypotheses and empirical evidence, and new research questions. *Ibis*, 148: 66–78.

Kalmbach E., van der Aa P., Komdeur J. (2005). Adoption as a gosling strategy to obtain better parental care? Experimental evidence for gosling choice and age-dependency of adoption in greylag geese. *Behaviour*, 142: 1515–33.

Kaminski J., Riedel J., Call J., et al. (2005). Domestic goats, *Capra hircus*, follow gaze direction and use social cues in an object choice task. *Animal Behaviour*, 69: 11–18.

Kampp K., Preuss N. O. (2005). The Greylag Geese of Utterslev Mose: a long-term population study of wild geese in an urban setting. *Dansk Ornithologisk Forenings Tidsskrift*, 99: 1–78.

Kanwisher J. W., Williams T. C., Teal J. M., et al. (1978). Radiotelemetry of heart rate from free-ranging gulls. *Auk*, 95: 288–93.

Kappeler P. M. (2010). *Animal Behaviour: Evolution and Mechanisms*. Springer, Heidelberg, Germany.

Kappeler P. M., Wimmer B., Zinner D., Tautz D. (2002). The hidden matrilineal structure of a solitary lemur: implications for primate social evolution. *Proceedings of the Royal Society B*, 269: 1755–63.

Keenleyside M. H. A. (1991). *Cichlid Fishes*. Chapman & Hall, London.

Kehmeier S., Schloegl C., Scheiber I. B. R., et al. (2011). Early development of gaze following into distant space in juvenile greylag geese (*Anser anser*). *Animal Cognition*, 14: 477–85.

Ketterson E. D., Nolan V., Jr, Sandell M. (2005). Testosterone in females: mediator of adaptive traits, constraint on sexual dimorphism, or both? *American Naturalist*, 166: 585–98.

Kikkawa J. (1980). Winter survival in relation to dominance classes among silvereyes *Zosterops lateralis chlorocephala* of Heron Island, Great Barrier Reef. *Ibis*, 122: 437–46.

Kikkawa J., Wilson J. M. (1983). Breeding and dominance among the Heron Island silvereyes *Zosterops lateralis chlorocephala*. *Emu*, 83: 181–98.

Kikkawa J., Smith J. N. M., Prys-Jones R., et al. (1986). Determinants of social dominance and inheritance of agonistic behavior in an island population of silvereyes, *Zosterops lateralis*. *Behavioral Ecology and Sociobiology*, 19: 165–9.

Kikuchi M., Yamaguchi N., Sato F., et al. (1994). Extraction methods for fecal hormone analysis in birds. *Journal of Ornithology*, 135: 64.

Klein R. M., Andrew R. J. (1986). Distraction, decisions and persistence in runway tests using the domestic chick. *Behaviour*, 99: 139–56.

Komdeur J., Hatchwell B. J. (1999). Kin recognition: function and mechanism in avian societies. *Trends in Ecology and Evolution*, 14: 237–41.

Koolhaas J. M., Korte S. M., De Boer S. F., et al. (1999). Coping styles in animals: current status in behavior and stress physiology. *Neuroscience and Biobehavioral Reviews*, 23: 925–35.

Koolhaas J. M., Korte S. M., De Boer S. F., et al. (2001). How and why coping systems vary among individuals. In Broom D. M. (ed.) *Coping with Challenge: Welfare in Animals Including Humans*. Dahlem University Press, Dahlem, Germany, pp. 197–209.

Korb J. (2010). Social insects, major evolutionary transitions and multilevel selection. *In* Kappeler P. M. (ed.) *Animal Behaviour: Evolution and Mechanisms.* Springer Verlag, Berlin, pp. 179–211.

Korte S. M., Beuving G., Ruesink W., *et al.* (1997). Plasma catecholamine and corticosterone levels during manual restraint in chicks from a high and low feather pecking line of laying hens. *Physiology and Behavior,* 62: 437–41.

Kotrschal K. (1995) *Im Egoismus vereint?* Piper Verlag, Munich.

Kotrschal K. (1998). Sensory systems. *In* Horn M, Martin K, Chotkowski M (eds.) *Intertidal Fishes.* Academic Press, San Diego, CA, pp. 126–42.

Kotrschal K., Hemetsberger J. (1995). Social constraints in feeding and individual specialization for transportable food in greylag geese (*Anser anser*). *Ökologie der Vögel,* 17: 157–63.

Kotrschal K., Palzenberger M. (1992). Neuroecology of cyprinids: comparative, quantitative histology reveals diverse brain patterns. *Environmental Biology of Fishes,* 33: 135–52.

Kotrschal K., Hemetsberger J., Dittami J. (1993). Food exploitation by a winter flock of greylag geese: behavioral dynamics, competition and social status. *Behavioral Ecology and Sociobiology,* 33: 289–95.

Kotrschal K., Hirschenhauser K., Möstl E. (1998a). The relationship between social stress and dominance is seasonal in greylag geese. *Animal Behaviour,* 55: 171–6.

Kotrschal K., Van Staaden M. J., Huber R. (1998b). Fish brains: evolution and environmental relationships. *Review in Fish Biology and Fisheries,* 8: 373–408.

Kotrschal K., Dittami J., Hirschenhauser K., *et al.* (2000). Effects of physiological and social challenges in different seasons on fecal testosterone and corticosterone in male domestic geese (*Anser domesticus*). *Acta Ethologica,* 2: 115–22.

Kotrschal K., Müller G., Winkler H. (2001). *Konzepte der Verhaltensforschung (Concepts of Ethology).* Filander Verlag, Fürth, Germany.

Kotrschal K., Hemetsberger J., Weiß B. M. (2006). Making the best of a bad situation: homosociality in male greylag geese. *In* Sommer V., Vasey P. (eds.) *Homosexual Behaviour in Animals: An Evolutionary Perspective.* Cambridge University Press, Cambridge, pp. 45–76.

Kotrschal K., Scheiber I. B. R., Hirschenhauser K. (2010). Individual performance in complex social systems: the greylag goose example. *In* Kappeler P. M. (ed.) *Animal Behaviour: Evolution and Mechanisms.* Springer Verlag, Berlin, pp. 121–48.

Kralj-Fišer S., Scheiber I. B. R., Blejec A., *et al.* (2007). Individualities in a flock of free-roaming greylag geese: behavioral and physiological consistency over time and across situations. *Hormones and Behavior,* 51: 239–48.

Kralj-Fišer S., Scheiber I. B. R., Kotrschal K., *et al.* (2010a). Glucocorticoids enhance and suppress heart rate and behaviour in time dependent manner in greylag geese (*Anser anser*). *Physiology and Behavior,* 100: 394–400.

Kralj-Fišer S., Weiß B. M., Kotrschal K. (2010b). Behavioural and physiological correlates of personality in greylag geese (*Anser anser*). *Journal of Ethology,* 28: 363–70.

Krause J., Ruxton G. D. (2002). *Living in Groups.* Oxford University Press, Oxford.

Krawany M. (1996). *Die Entwicklung einer nicht invasiven Methode zum Nachweis von Steroidhormonmetaboliten im Kot von Gänsen.* Dissertation, Institute of Biochemistry, University of Veterinary Medicine, Vienna.

Kristiansen J. N. (1998). Nest predation in reedbed nesting greylag geese *Anser anser* in Vejlerne, Denmark. *Ardea,* 86: 137–45.

Kruckenberg H. (2005). Wann werden 'die Kleinen' endlich erwachsen? Untersuchungen zum Familienzusammenhalt farbmarkierter Blessgänse *Anser albifrons albifrons*. *Vogelwelt*, 126: 253–8.

Kruuk L. E. B. (2004). Estimating genetic parameters in natural populations using the 'animal' model. *Philosophical Transactions of the Royal Society B*, 359: 873–90.

Kutsukake N. (2011). Book Review: Peter Kappeler (ed.): Animal behaviour: evolution and mechanisms. *Primates*, 52: 93–5.

Kutsukake N., Clutton-Brock T. (2008). Do meerkats engage in conflict management following aggression? Reconciliation, submission and avoidance. *Animal Behaviour*, 75: 1441–53.

Lack D. (1968). *Ecological Adaptations for Breeding in Birds*. Methuen, London.

Lamprecht J. (1977). A comparison of the attachment to parents and siblings in juvenile geese (*Branta canadensis* and *Anser indicus*). *Zeitschrift für Tierpsychologie*, 43: 415–24.

Lamprecht J. (1985). Dominanz und Fortpflanzungserfolg bei Streifengänsen (*Anser indicus*): eine multiple Regressionsanalyse. *Journal of Ornithology*, 126: 287–93.

Lamprecht J. (1986a). Structure and causation of the dominance hierarchy in a flock of Bar-headed Geese (*Anser indicus*). *Behaviour*, 96: 28–48.

Lamprecht J. (1986b). Social dominance and reproductive success in a goose flock (*Anser indicus*). *Behaviour*, 97: 50–65.

Lamprecht J. (1987). Female reproductive strategies in bar-headed geese (*Anser indicus*): why are geese monogamous? *Behavioral Ecology and Sociobiology*, 21: 297–305.

Lamprecht J. (1990). Predicting current reproductive success of goose pairs *Anser indicus* from male and female reproductive history. *Ethology*, 85: 123–31.

Lamprecht J. (1991). Factors influencing leadership: a study of goose families (*Anser indicus*). *Ethology*, 89: 265–74.

Lamprecht J. (1992). Variable leadership in barheaded geese (*Anser indicus*): an analysis of pair and family departures. *Behaviour*, 122: 105–20.

Langen T. A. (2000). Prolonged offspring dependence and cooperative breeding in birds. *Behavioral Ecology*, 11: 367–77.

Lank D. B., Rockwell R. F., Cooke F. (1990). Frequency-dependent fitness consequences of intraspecific nest parasitism in snow geese. *Evolution*, 44: 1436–53.

Larsson K., Tegelström H., Forslund P. (1995). Intraspecific nest parasitism and adoption of young in the barnacle goose: effects of survival and reproductive performance. *Animal Behaviour*, 50: 1349–60.

Lazareva O. F., Wasserman E. A. (2006). Effect of stimulus orderability and reinforcement history on transitive responding in pigeons. *Behavioural Processes*, 72: 161–72.

Lazareva O. F., Smirnova A. A., Bagozkaja M. S., et al. (2004). Transitive responding in hooded crows requires linearly ordered stimuli. *Journal of the Experimental Analysis of Behavior*, 82: 1–19.

Lazarus J. (1978). Vigilance, flock size, and domain of danger size in the white-fronted goose. *Wildfowl*, 29: 135–45.

Lessells C. M., Boag P. T. (1987). Unrepeatable repeatabilities: a common mistake. *Auk*, 104: 116–21.

Levine S. (1993a). The influence of social factors on the response of stress. *Psychotherapy and Psychosomatics*, 60: 33–8.

Levine S. (1993b). The psychoendocrinology of stress. *Annals of the New York Academy of Sciences*, 697: 61–9.

Livezey B. C. (1996). A phylogenetic analysis of geese and swans (Anseriformes: Anserinae), including selected fossil species. *Systematic Biology*, 45: 415–50.

Loery G., Pollock K. H., Nichols J. D., *et al.* (1987). Age-specificity of black-capped chickadee survival rates: analysis of capture–recapture data. *Ecology*, 68: 1038–44.

Lonsdorf E. V., Ross S. R., Linick S. A., *et al.* (2009). An experimental, comparative investigation of tool use in chimpanzees and gorillas. *Animal Behaviour*, 77: 1119–26.

Loonen M. J. J. (1997). *Goose breeding ecology: overcoming successive hurdles to raise goslings.* Dissertation. Mathematics and Natural Sciences, University of Groningen, Netherlands.

Loonen M. J. J., Bruinzeel L. W., Black J. M., *et al.* (1999). The benefit of large broods in barnacle geese: a study using natural and experimental manipulations. *Journal of Animal Ecology*, 68: 753–68.

Lorenz K. (1935). Der Kumpan in der Umwelt des Vogels. *Journal für Ornithologie*, 83: 137–213, 289–413.

Lorenz K. (1939a). Vergleichende Verhaltensforschung. *Zoologischer Anzeiger*, Supplement 12: 69–102.

Lorenz K. (1939b). Die Paarbildung beim Kolkraben. *Zeitschrift für Tierpsychologie*, 3: 278–92.

Lorenz K. (1940). Durch Domestikation verursachte Störungen arteigenen Verhaltens. *Zeitschrift für angewandte Psychologie und Charakterkunde* 59: 2–81.

Lorenz K. (1941). Vergleichende Bewegungsstudien von Anatiden. *Journal für Ornithologie*, 89: 194–294.

Lorenz K. (1963). *Das sogenannte Böse: Zur Naturgeschichte der Aggression.* Verlag Dr. G. Borotha-Schoeler, Vienna.

Lorenz K. (1973). *Die acht Todsünden der zivilisierten Menschheit.* Piper Verlag, Munich.

Lorenz K. (1988). *Hier Bin Ich: Wo Bist Du? Ethologie der Graugans.* Piper Verlag, Munich.

Lorenz K. (1991). *Here I Am: Where Are You?* Harcourt Brace Jovanovich, New York.

Lorenz K. (1992). *Die Naturwissenschaft vom Menschen: Eine Einführung in die vergleichende Verhaltensforschung. Das 'Russische Manuskript'.* Piper Verlag, Munich.

Lorenz K., Tinbergen N. (1938). Taxis und Instinkthandlung in der Eirollbewegung der Graugans I. *Zeitschrift für Tierpsychologie*, 2: 1–29.

Lorenz K., Kalas S., Kalas K. (1978). *Das Jahr der Graugans.* Piper Verlag, Munich.

Lorenz K., Kalas S., Kalas K. (1979). *The Year of the Greylag Goose.* Harcourt Brace Jovanovich, San Diego, CA.

Loretto M.-C., Schloegl C., Bugynar T. (2010). Northern bald ibis follow others' gaze into distant space but not behind barriers. *Biology Letters*, 6: 14–17.

Lyon B. E., Eadie J. M. (2008). Conspecific brood parasitism in birds: a life-history perspective. *Annual Review of Ecology, Evolution and Systematics*, 39: 343–63.

MacArthur R. A., Geist V., Johnston R. H. (1982). Cardiac and behavioral responses of mountain sheep to human disturbance. *Journal of Wildlife Management*, 46: 351–8.

Machlis L., Dodd P. W. D., Fentres J. C. (1985). The pooling fallacy: problems arising when individuals contribute more than one observation to the data set. *Zeitschrift für Tierpsychologie*, 68: 201–14.

MacLean E. L., Merritt D. J., Brannon E. M. (2008). Social complexity predicts transitive reasoning in prosimian primates. *Animal Behaviour*, 76: 479–86.

Macphail E. M. (2001). Conservation in the neurology and psychology of cognition in vertebrates. *In:* Roth G., Wullimann M. F. (eds.) *Brain, Evolution and Cognition.* John Wiley & Sons, New York, pp. 401–30.

Madge S., Burn H. (1988). *Waterfowl: An Identification Guide to Ducks, Geese and Swans of the World.* Houghton Mifflin, Boston, MA.

Madsen J., Cracknell G., Fox A. D. (1999). *Goose Populations of the Western Palearctic: A Review of Status and Distribution.* National Environmental Research Institute, Rönde, Denmark.

Magnusson M. S. (2000). Discovering hidden time patterns in behavior: T-patterns and their detection. *Behavior Research Methods, Instruments and Computers*, 32: 93–110.

Marchetti C., Drent P. J. (2000). Individual differences in the use of social information in foraging by captive great tits. *Animal Behaviour*, 60: 131–40.

Marino L. (1996). What can dolphins tell us about primate evolution? *Evolutionary Anthropology*, 5: 81–6.

Martin K. (1995). Patterns and mechanisms for age-dependent reproduction and survival in birds. *American Zoologist*, 35: 340–8.

Mausz B., Dittami J., Kotrschal K. (1992). Triumphgeschrei und Aggression bei der Graugans (*Anser anser*). *Ökologie der Vögel (Ecology of Birds)*, 14: 165–72.

McComb K., Moss C. J., Durant S. M., *et al.* (2001). Matriarchs as repositories of social knowledge in African elephants. *Science*, 292: 491–4.

McCraty R., Atkinson M., Tiller T. C., *et al.* (1995). The effects of emotions on short-term power spectrum analysis of heart rate variability. *American Journal of Cardiology*, 76: 1089–93.

McEwen B., Wingfield J. C. (2003). The concept of allostasis in biology and biomedicine. *Hormones and Behavior*, 43: 2–15.

McGonigle B. O., Chalmers M. (1977). Are monkeys logical? *Nature*, 267: 694–6.

McNamara J. M., Forslund P., Lang A. (1999). An ESS model for divorce strategies in birds. *Philosophical Transactions of the Royal Society B.*, 354: 223–36.

Mendl M., Deag J. M. (1995). How useful are the concepts of alternative strategy and coping strategy in applied studies of social behaviour? *Applied Animal Behaviour Science*, 44: 119–37.

Meyers B. (1932). *Meyers Blitz-Lexikon: Die Schnellauskunft für Jedermann in Wort und Bild.* Bibliographisches Institut A.G., Leipzig, Germany.

Milinski M. (1987). Tit for tat in sticklebacks and the evolution of cooperation. *Nature*, 325: 433–5.

Mitchell C., Colhoun K., Fox A. D., *et al.* (2010). Trends in goose numbers wintering in Britain and Ireland, 1995 to 2008. *Ornis Svecica*, 20: 128–43.

Moll H., Tomasello M. (2007). Cooperation and human cognition: the Vygotskian intelligence hypothesis. *Philosophical Transactions of the Royal Society B*, 362: 639–48.

Moore M. C. (1982). Hormonal responses of free-living male white-crowned sparrows to experimental manipulations of female sexual behavior. *Hormones and Behavior*, 16: 323–9.

Morris D. M., Sekyra F., Glick B. (1971). Discrimination performance in male chicks treated embryonically with testosterone. *Poultry Science*, 50: 289–91.

Möstl E., Meyer H. H. D., Bamberg E., *et al.* (1987). Oestrogen determination in feces of mares by enzyme immunoassay on microtitre plates. *In Proceedings of the Symposium on Analysis Steroids*, Sopron, Hungary, pp. 219–24.

Möstl E., Rettenbacher S., Palme R. (2005). Measurement of corticosterone metabolites in birds' droppings: an analytical approach. *Annals of the New York Academy of Sciences*, 1046: 17–34.

Mulder R. A., Williams T. D., Cooke F. (1995). Dominance, brood size and foraging behavior during brood-rearing in the lesser snow goose: an experimental study. *Condor*, 97: 99–106.

Naguib M., Kazek A., Schaper S. V., *et al.* (2010). Singing activity reveals personality traits in great tits. *Ethology*, 116: 763–9.

Nathan R., Spiegel O., Fortmann-Roe S., *et al.* (2012). Using tri-axial acceleration data to identify behavioral modes of free-ranging animals: general concepts and tools illustrated for griffon vultures. *Journal of Experimental Biology*, 215: 986–96.

Naves L. C., Cam E., Monnat J.-Y. (2007). Pair duration, breeding success and divorce in a long-lived sea-bird: benefits of mate familiarity? *Animal Behaviour*, 73: 433–44.

Nealen P. M., Ricklefs R. E. (2001). Early diversification of the avian brain: body relationship. *Journal of Zoology*, 253: 391–404.

Neff B. D., Pitcher T. E. (2005). Genetic quality and sexual selection: an integrated framework for good genes and compatible genes. *Molecular Ecology*, 14: 19–38.

Nephew B. C., Romero L. M. (2003). Behavioral, physiological, and endocrine responses of starlings to acute increases in density. *Hormones and Behavior*, 44: 222–32.

Nephew B. C., Kahn S. A., Romero L. M. (2003). Heart rate and behavior are regulated independently of corticosterone following diverse acute stressors. *General and Comparative Endocrinology*, 133: 173–80.

Netto W. J., van Hooff J. A. R. A. M. (1986). Conflict interference and the development of dominance relationships in immature *Macaca fascicularis*. In Else J. G., Lee P. C. (eds.) *Primate Ontogeny: Cognition and Social Behaviour*. Cambridge University Press, New York, pp. 291–300.

Newton I., Kerbes R. H. (1974). Breeding of greylag geese (*Anser anser*) on the Outer Hebrides, Scotland. *Journal of Animal Ecology*, 43: 771–83.

Nilsson J. A. (1990). Family break-up: spontaneous dispersal or parental aggression? *Animal Behaviour*, 40: 1001–3.

Nilsson L., Persson H. (1994). Factors affecting the breeding performance of a marked Greylag Goose *Anser anser* population in south Sweden. *Wildfowl*, 45: 33–48.

Nilsson L., Persson H. (1996). The influence of the choice of winter quarters on the survival and breeding performance of greylag geese (*Anser anser*). In Birkan M., van Vessem J., Havet P., *et al.* (eds.) *Proceedings of the Anatidae 2000 Conference*, Vol. 13. Game Wildlife, Strasbourg, France, pp. 557–71.

Nol E., Cheng K., Nichols C. (1996). Heritability and phenotypic correlations of behaviour and dominance rank of Japanese quail. *Animal Behaviour*, 52: 813–20.

Obrist P. A. (1981). *Cardiovascular Psychophysiology*. Plenum Press, New York.

Ogilvie M., Pearson B. (1994). *Wildfowl*. Hamlyn, London.

O'Keefe J., Nadel L. (1978). *The Hippocampus as a Cognitive Map*. Oxford University Press, Oxford.

Oliveira R. F., McGregor P. K., Latruffe C. (1998). Know thine enemy: fighting fish gather information from observing conspecific interactions. *Proceedings of the Royal Society B*, 265: 1045–9.

Orians, G. H. (1969). On the evolution of mating systems in birds and mammals. *American Naturalist*, 103: 589–603.

Osiejuk T. S., Kuczynski L. (2007). Factors affecting flushing distance in incubating female greylag geese *Anser anser*. *Wildlife Biology*, 13: 11–18.

Osvath M., Osvath H. (2008). Chimpanzee (*Pan troglodytes*) and orangutan (*Pongo abelii*) forethought: self-control and pre-experience in the face of future tool use. *Animal Cognition*, 11: 661–74.

Otter K., Ratcliffe L. M., Michaud D., *et al.* (1998). Do female black-capped chickadees prefer high-ranking males as extra-pair partners? *Behavioral Ecology and Sociobiology*, 43: 25–35.

Øverli O., Sørensen C., Pulman K. G. T., *et al.* (2007). Evolutionary background for stress-coping styles: relationships between physiological, behavioral, and cognitive traits in non-mammalian vertebrates. *Neuroscience and Biobehavioral Reviews*, 31: 396–412.

Owen M. (1980). *Wild Geese of the World*. Batsford Press, London.

Owen M. (1990). The barnacle goose. *Shire Natural History*, 51: 1–23.

Owen M., Black J. M. (1989). Factors affecting the survival of barnacle geese on migration from the breeding grounds. *Journal of Animal Ecology*, 58: 603–17.

Palme R., Möstl E. (1993). Biotin-strepavidin enzyme immunoassay for the determination of estrogens and androgens in boar feces. In Görög S. (ed.) *Advances of Steroid Analysis*. Akademie Kiado, Budapest, Hungary, pp. 111–17.

Panksepp J. (1998). *Affective Neuroscience: The Foundations of Human and Animal Emotions*. Oxford University Press, New York.

Paradis E., Bailie S. R., Sutherland W. J., *et al.* (2000). Spatial synchrony in populations of birds: effects of habitat, population trend, and spatial scale. *Ecology*, 81: 2112–25.

Parr L. A., Winslow J. T., Hopkins W. D., *et al.* (2000). Recognizing facial cues: individual discrimination by chimpanzees (*Pan troglodytes*) and rhesus monkeys (*Macaca mulatta*). *Journal of Comparative Psychology*, 114: 47–60.

Paz-y-Miño G., Bond A. B., Kamil A. C., *et al.* (2004). Pinyon jays use transitive inference to predict social dominance. *Nature*, 430: 778–81.

Penn D. C., Povinelli D. J. (2007). On the lack of evidence that non-human animals possess anything remotely resembling a 'theory of mind'. *Philosophical Transactions of the Royal Society B*, 362: 731–44.

Pepperberg I. M. (1999). *The Alex Studies: Cognitive and Communicative Abilities of Grey Parrots*. Harvard University Press, Cambridge, MA.

Pereira M. E. (1992). The development of dominance relations before puberty in cercopithecine societies. In Silverberg J., Gray J. P. (eds.) *Aggression and Peacefulness in Humans and Other Primates*. Oxford University Press, Oxford, pp. 117–49.

Perfito N., Schirato G., Brown M., *et al.* (2002). Response to acute stress in the Harlequin Duck (*Histrionicus histrionicus*) during the breeding season and moult: relationships to gender, condition, and life-history stage. *Canadian Journal of Zoology*, 80: 1334–43.

Perry S., Manson J. H., Muniz L., Gros-Louis J., Vigilant L. (2008). Kin-biased social behaviour in wild adult female white-faced capuchins, *Cebus capucinus*. *Animal Behaviour*, 76: 187–99.

Pfeffer K., Fritz J., Kotrschal K. (2002). Hormonal correlates of being an innovative greylag goose, *Anser anser*. *Animal Behaviour*, 63: 687–95.

Pistorius P. A., Follestad A., Nilsson L., *et al.* (2007). A demographic comparison of two Nordic populations of Greylag Geese *Anser anser*. *Ibis*, 149: 553–63.

Poisbleau M., Fritz H., Valeix M., *et al.* (2006). Social dominance correlates and family status in wintering dark-bellied brent geese, *Branta bernicla bernicla*. *Animal Behaviour*, 71: 1351–8.

Povinelli D. J., Eddy T. J. (1996). Chimpanzees: joint visual attention. *Psychological Science*, 7: 129–35.

Pöysä H., Pesonen M. (2007). Nest predation and the evolution of conspecific brood parasitism: from risk spreading to risk assessment. *American Naturalist*, 169: 94–104.

Pratt L. (1994). *Odyssey* 19.535–50: On the interpretation of dreams and signs in Homer. *Classical Philology*, 89: 147–52.

Preston S. D., de Waal F. B. M. (2002). Empathy: its ultimate and proximate bases. *Behavioral and Brain Sciences*, 25: 1–20.

Prevett J. P., MacInnes C. D. (1980). Family and other social groups in snow geese. *Wildlife Monographs*, 71: 1–46.

Prinzinger R., Nagel B., Bahat O., *et al.* (2002). Energy metabolism and body temperature in the griffon vulture (*Gyps fulvus*) with comparative data on the hooded vulture (*Necrosyrtes monachus*) and the white-backed vulture (*Gyps africanus*). *Journal of Ornithology*, 143: 456–67.

Prop J., Deerenberg C. (1991). Spring staging in Brent geese *Branta bernicla*: feeding constraints and the impact of diet on the accumulation of body reserves. *Oecologia*, 87: 19–28.

Prop J., van Eerden M. R., Drent R. H. (1984). Reproductive success in the barnacle goose *Branta leucopsis* in relation to food exploitation on the breeding grounds, western Spitsbergen. *Norsk Polarinstitutt Skrifter*, 181: 87–117.

Pusey A., Williams J., Goodall J. (1997). The influence of dominance rank on the reproductive success of female chimpanzees. *Science*, 277: 828–31.

Raby C. R., Alexis D. M., Dickinson A., *et al.* (2007). Planning for the future by Western scrub-jays. *Nature*, 445: 919–21.

Radesäter T. (1974). Form and sequential associations between the triumph ceremony and other behaviour patterns in the Canada goose *Branta canadensis* L. *Ornis Scandinavica*, 5: 87–101.

Ranta E., Rita H., Lindström K. (1993). Competition versus cooperation: success of individuals foraging alone and in groups. *American Naturalist*, 142: 42–58.

Raveling D. G. (1969a). Preflight and flight behavior of Canada geese. *Auk*, 86: 671–81.

Raveling D. G. (1969b). Social classes of Canada geese in winter. *Journal of Wildlife Management*, 33: 304–18.

Raveling D. G. (1970). Dominance relationships and agonistic behavior of Canada geese in winter. *Behaviour*, 37: 291–319.

Raveling D. G., Sedinger J. S., Johnsson D. S. (2000). Reproductive success and survival in relation to experience during the first two years in Canada geese. *Condor*, 102: 941–5.

Réale D., Gallant B. Y., Leblanc M., *et al.* (2000). Consistency in temperament in bighorn ewes and correlates with behaviour and life history. *Animal Behaviour*, 60: 589–97.

Réale D., Reader S. M., Sol D., *et al.* (2007). Integrating animal temperament within ecology and evolution. *Biological Reviews*, 82: 291–318.

Reichart L. M., Anderholm S., Muñoz-Fuentes V., *et al.* (2010). Molecular identification of brood-parasitic females reveals an opportunistic reproductive tactic in ruddy ducks. *Molecular Ecology*, 19: 401–13.

Remage-Healey L., Adkins-Regan E., Romero L. M. (2003). Behavioral and adrenocortical responses to mate separation and reunion in the zebra finch. *Hormones and Behavior*, 43: 108–14.

Richner H. (1989). Phenotypic correlates of dominance in carrion crows and their effects on access to food. *Animal Behaviour*, 38: 606–12.

Riedstra B. J., Groothuis T. G. G. (2000). The effects of prenatal androgen treatment in domestic chickens. *In* 7th International Symposium on Avian Endocrinology, Varanasi, India.

Rödel H. G., Monclús R., von Holst D. (2006). Behavioral styles in European rabbits: social interactions and responses to experimental stressors. *Physiology and Behavior*, 89: 180–8.

Roper T. J. (1999). Olfaction in birds. *Advances in the Study of Behaviour*, 28: 247–332.

Rost W. (2001). *Emotionen: Elixiere des Lebens*, Vol. 2. Springer Verlag, Heidelberg, Germany.

Rowley I. (1983). Re-mating in birds. *In* Bateson P. (ed.) *Mate Choice*. Cambridge University Press, Cambridge, pp. 331–60.

Rubenstein D. I. (1978). On predation, competition and the advantages of group living. *In* Bateson P. G., Klopfer P. H. (eds.) *Perspectives in Ethology*. Plenum Press, New York, pp. 205–31.

Ruckstuhl K. E. (1999). To synchronize or not to synchronize: a dilemma for young bighorn males? *Behaviour*, 136: 805–18.

Ruis M. A. W., Brake J. H. A., van de Burgwal J. A., *et al.* (2000). Personalities in female domesticated pigs: behavioural and physiological indications. *Applied Animal Behaviour Science*, 66: 31–47.

Ruis M. A. W., Brake J. H. A., Engel B., *et al.* (2001). Adaptation to social isolation: acute and long-term stress responses of growing gilts with different coping characteristics. *Physiology and Behavior*, 73: 541–51.

Rutschke E. (1982). Stability and dynamics in the social structure of the Greylag Goose (*Anser anser*). *Aquila*, 89: 39–55.

Rutschke E. (1997). *Die Wildgänse Europas*. Deutscher Landwirtschaftsverlag, Berlin.

Sachser N., Dürschlag M., Hirzel D. (1998). Social relationships and the management of stress. *Psychoneuroendocrinology*, 23: 891–904.

Sakamoto K. Q., Sato K., Ishizuka M., *et al.* (2009). Can ethograms be automatically generated using body acceleration data from free-ranging birds? *PloS One*, 4: e5379.

Samelius G., Alisauskas R. T. (2001). Deterring arctic fox predation: the role of parental nest attendance by lesser snow geese. *Canadian Journal of Zoology*, 79: 861–6.

Sandi C., Rose S. P. R. (1994). Corticosterone enhances long-term retention in one day-old chicks trained in a weak passive avoidance learning paradigm. *Brain Research*, 647: 106–12.

Sapolsky R. M. (1992). Neuroendocrinology of the stress response. *In* Becker J., Breedlove S., Crews D. (eds.) *Behavioural Endocrinology*. MIT University Press, Cambridge, MA.

Sapolsky R. M. (1993). Endocrinology alfresco: psychoendocrine studies of wild baboons. *Recent Progress in Hormone Research*, 48: 437–68.

Sapolsky R. M. (1994). *Why Zebras Don't Get Ulcers: An Updated Guide to Stress, Stress-Related Disease and Coping*. Freeman and Co., New York.

Sapolsky R. M. (1997). The importance of a well-groomed child. *Science*, 277: 1620–1.

Sapolsky R. M. (2005). The influence of social hierarchy on primate health. *Science*, 308: 648–52.

Sapolsky R. M., Romero L. M., Munck A. U. (2000). How do glucocorticoids influence stress responses? Integrating permissive, suppressive, stimulatory, and preparative actions. *Endocrine Reviews*, 21: 55–89.

Sargeant A. B., Raveling D. G. (1992). Mortality during the breeding season. *In* Batt B. D. J., Afton A. D., Anderson M. G., *et al.* (eds.) *Ecology and Management of Breeding Waterfowl*. The University of Minnesota Press, Minneapolis, MN, pp. 396–422.

Scheiber I. B. R., Weiß B. M., Frigerio D., *et al.* (2005a). Active and passive social support in families of greylag geese (*Anser anser*). *Behaviour*, 142: 1535–57.

Scheiber I. B. R., Kralj S., Kotrschal K. (2005b). Sampling effort/frequency necessary to infer individual acute stress responses from fecal analysis in greylag geese (*Anser anser*). *Annals of the New York Academy of Sciences*, 1046: 154–67.

Scheiber I. B. R., Weiß B. M., Hirschenhauser K., *et al.* (2008). Does 'relationship intelligence' make big brains in birds? *The Open Access Biology Journal*, 1: 6–8.

Scheiber I. B. R., Kotrschal K., Weiß B. M. (2009a). Benefits of family reunions: social support in secondary greylag goose families. *Hormones and Behavior*, 55: 133–8.

Scheiber I. B. R., Kotrschal K., Weiß B. M. (2009b). Serial agonistic attacks by greylag goose families (*Anser anser*) against the same opponent. *Animal Behaviour*, 77: 1211–16.

Scheiber I. B. R., Hohnstein A., Kotrschal K., *et al.* (2011). Juvenile greylag geese (*Anser anser*) discriminate between individual siblings. *PLoS One*, 6(8): e22853.

Scherer K. R. (2005). What are emotions? And how can they be measured? *Trends and Developments: Research on Emotions*, 44: 695–729.

Schino G. (2000). Beyond the primates. *In* Aureli F., De Waal F. B. M. (eds.) *Natural Conflict Resolution*. University of California Press, Berkeley, CA, pp. 225–42.

Schloegl C., Kotrschal K., Bugnyar T. (2007). Gaze following in common ravens, *Corvus corax*: ontogeny and habituation. *Animal Behaviour*, 74: 769–78.

Schloegl C., Schmidt J., Scheid C., *et al.* (2008). Gaze-following in non-human animals: the corvid example. *In* Columbus F. (ed.) *Animal Behaviour: New Research*. Nova Science Publishers, New York.

Schmutz J. A. (1993). Survival and pre-fledging body mass in juvenile emperor geese. *Condor*, 95: 222–5.

Schneider J. M., Lamprecht J. (1990). The importance of biparental care in a precocial, monogamous bird, the bar-headed goose (*Anser indicus*). *Behavioral Ecology and Sociobiology*, 27: 415–19.

Schutz F. (1965). Sexuelle Prägung bei Anatiden. *Zeitschrift für Tierpsychologie*, 22: 50–103.

Schütz K., Wallner B., Kotrschal K. (1997). Diurnal patters of steroid hormones from feces in greylag goslings (*Anser anser*). *Advances in Ethology*, 32: 66.

Schwabl H. (1993). Yolk is a source of maternal testosterone for developing birds. *Proceedings of the National Academy of Sciences USA*, 90: 11446–50.

Schwabl H. (1996a). Maternal testosterone in the avian egg enhances postnatal growth. *Comparative Biochemistry and Physiology*, 114A: 271–6.

Schwabl H. (1996b). Environment modifies the testosterone levels of a female bird and its eggs. *Journal of Experimental Zoology*, 276: 157–63.

Schwabl H. (1997). Maternal steroid hormones in the egg. *In* Harvey S., Etches R. J. (eds.) *Perspectives in Avian Endocrinology*. Journal of Endocrinology Ltd, Bristol, UK, pp. 3–13.

Schwagmeyer P. L. (1980). Alarm calling behavior of the thirteen-lined ground squirrel, *Spermophilus tridecemlineatus*. *Behavioral Ecology and Sociobiology*, 7: 195–200.

Scott D. K. (1980). Functional aspects of prolonged parental care in Bewick's swans. *Animal Behaviour*, 28: 938–52.

Seed A. M., Clayton N. S., Emery N. J. (2008). Cooperative problem solving in rooks (*Corvus frugilegus*). *Proceedings of the Royal Society B*, 275: 1421–9.

Segelbacher G., Höglund J. (2009). Ecological genomics. *Genetica*, 136: 387–90.

Selye H. (1950). *The Physiology and Pathology of Exposure to Stress*. ACTA Inc. Medical Publishers, Montreal, Canada.

Senar J. C. (1984). Allofeeding in Eurasian siskins (*Carduelis spinus*). *Condor*, 86: 213–14.

Sgoifo A., Boer S. F., Haller J., *et al.* (1996). Individual differences in plasma catecholamine and corticosterone stress response in wildtype rats: relationship with aggression. *Physiology and Behavior*, 60: 1403–7.

Sgoifo A., Costoli T., Meerlo P., *et al.* (2005). Individual differences in cardiovascular response to social challenge. *Neuroscience and Biobehavioral Reviews*, 29: 59–66.

Shettleworth S. J. (2010). *Cognition, Evolution and Behavior*, 2nd edn. Oxford University Press, New York.

Shields W. M. (1984). Barn swallow mobbing: self-defence, collateral kin defence, group defence or parental care? *Animal Behaviour*, 32: 132–48.

Shultz S., Dunbar R. I. M. (2006). Both social and ecological factors predict ungulate brain size. *Proceedings of the Royal Society B*, 273: 207–15.

Shultz S., Dunbar R. I. M. (2007). The evolution of the social brain: anthropoid primates contrast with other vertebrates. *Proceedings of the Royal Society B*, 274: 2429–36.

Sibley C. G., Monroe B. L., Jr (1990). *Distribution and Taxonomy of Birds of the World*. Yale University Press, New Haven, CT.

Siegel H. S. (1980). Physiological stress in birds. *BioScience*, 30: 529–33.

Sih A., Bell A. M., Johnson J. C. (2004). Behavioral syndromes: an ecological and evolutionary overview. *Trends in Ecology and Evolution*, 19: 372–8.

Silverin B. (1990). Testosterone and corticosterone and their relation to territorial and parental behavior in the pied flycatcher. In Balthazart J. (ed.) *Hormones, Brain and Behavior in Vertebrates. 2. Behavioural Activation in Males and Females: Social Interaction and Reproductive Endocrinology*. Karger Verlag, Basel, Switzerland, pp. 129–42.

Smith B. R., Blumstein D. T. (2008). Fitness consequences of personality: a meta-analysis. *Behavioral Ecology*, 19: 448–55.

Smith J. E., Van Horn R. C., Powning K. S., *et al.* (2010). Evolutionary forces favoring intragroup coalitions among spotted hyenas and other animals. *Behavioral Ecology*, 21: 284–303.

Smith J. E., Powning K. S., Dawes S. E., *et al.* (2011). Greetings promote cooperation and reinforce social bonds among spotted hyaenas. *Animal Behaviour*, 81: 401–15.

Smith S. F. (1978). Alarm calls, their origin and use in *Eutamias sonomae*. *Journal of Mammalogy*, 59: 888–93.

Smith T. E., McGreer-Whitworth B., French J. (1998). Close proximity of the heterosexual partner reduces the physiological and behavioral consequences of novel-cage housing in black tufted-ear marmosets (*Callithrix kuhli*). *Hormones and Behavior*, 34: 211–22.

Snow D. W. (1977). Duetting and other synchronized displays of the blue-backed manakin, *Chiroxiphia* spp. In Stonehouse B., Perrins C. M. (eds.) *Evolutionary Ecology*. MacMillan Press, London, pp. 239–51.

Soares M., Bshary R., Fusani L., *et al.* (2010). Hormonal mechanisms of cooperative behaviour. *Philosophical Transactions of the Royal Society B*, 365: 2737–50.

Sockman K. W., Schwabl H. (2000). Yolk androgens reduce offspring survival. *Proceedings of the Royal Society B*, 267: 1451–6.

Sommer V., Vasey P. L. (2006). *Homosexual Behaviour in Animals*. Cambridge University Press, Cambridge.

Sovrano V. A., Bisazza A., Vallortigara G. (2007). How fish do geometry in large and in small spaces. *Animal Cognition*, 10: 47–54.

Spoon T. R., Millam J. R., Owings D. H. (2004). Variation in the stability of cockatiel (*Nymphicus hollandicus*) pair relationships: the roles of males, females and mate compatibility. *Behaviour*, 141: 1211–34.

Spoon T. R., Millam J. R., Owings D. H. (2006). The importance of mate behavioural compatibility in parenting and reproductive success by cockatiels, *Nymphicus hollandicus*. *Animal Behaviour*, 71: 315–26.

Spoon T. R., Millam J. R., Owings D. H. (2007). Behavioural compatibility, extrapair copulation and mate switching in a socially monogamous parrot. *Animal Behaviour*, 73: 815–24.

Stahl J., Tolsma P., Loonen M. J. J., et al. (2001). Subordinates explore but dominants profit: resource competition in high Arctic barnacle goose flocks. *Animal Behaviour*, 61: 257–64.

Stamps J. A., Groothuis T. G. G. (2010a). The development of animal personality: relevance, concepts and perspectives. *Biological Reviews*, 85: 301–25.

Stamps J. A., Groothuis T. G. G. (2010b). Developmental perspectives on personality: implications for ecological and evolutionary studies of individual differences. *Philosophical Transactions of the Royal Society B*, 365: 4029–41.

Starck J. M., Ricklefs R. E. (1998). Patterns of development: the altricial-precocial spectrum. In Starck J. M., Ricklefs R. E. (eds.) *Avian Growth and Development: Evolution Within the Altricial Precocial Spectrum*. Oxford University Press, New York, pp. 1–28.

Steiger S. S., Fidler A. E., Valcu M., et al. (2008). Avian olfactory receptor gene repertoires: evidence for a well-developed sense of smell in birds? *Proceedings of the Royal Society B*, 275: 2309–17.

Stoddard S. L., Bergdall V. K., Townsend D., et al. (1986). Plasma catecholamines associated with hypothalamically elicited flight behavior. *Folia Primatologica*, 17: 709–15.

Summers C. H. (2002). Social interaction over time, implications for stress responsiveness. *Integrative and Comparative Biology*, 42: 591–9.

Summers C. H., Watt M. J., Ling T. L., et al. (2005). Glucocorticoid interaction with aggression in non-mammalian vertebrates: reciprocal action. *European Journal of Pharmacology*, 526: 21–35.

Swoboda R. (2006). *Social modulation of immuno-reactive corticosterone metabolites in goslings from hatching to fledging: ontogeny of social support in greylag geese (*Anser anser*)*. Diploma thesis, Department of Behavioural Biology, University of Vienna.

Taborsky M. (1984). Broodcare helpers in the cichlid fish, *Lamprologus brichardi*: their costs and benefits. *Animal Behaviour*, 32: 1236–52.

Tarvin K. A., Woolfenden G. E. (1997). Patterns of dominance and aggressive behavior in blue jays at a feeder. *Condor*, 99: 434–44.

Tatoyan S. K., Cherkovich G. M. (1972). The heart rate in monkeys (baboons and macaques) in different physiological states recorded by radiotelemetry. *Folia Primatologica*, 17: 255–66.

Taylor, S. E., Klein, L. C., Lewis, B. P., et al. (2000). Biobehavioral responses to stress in females: tend-and-befriend, not fight-or-flight. *Psychological Review*, 10: 411–29.

Taylor S. S., Leonard M. L., Boness D. J. (2001). Aggressive nest intrusions by male Humboldt penguins. *Condor*, 103: 162–5.

Tibbetts E. A. (2002). Visual signals of individual identity in the wasp *Polistes fuscatus*. *Proceedings of the Royal Society B*, 269: 1423–8.

Tinbergen N. (1963). On aims and methods of ethology. *Zeitschrift für Tierpsychologie*, 22: 50–103.

Todd F. S. (1996). *Natural History of the Waterfowl*. Ibis Publishing Company, Vista, CA.

Tolman E. C. (1948). Cognitive maps in rats and men. *Psychological Review*, 55: 189–208.

Tóth Z., Bókony V., Lendvai A. Z., *et al.* (2009). Whom do the sparrows follow? The effect of kinship on social preference in house sparrow flocks. *Behavioural Processes*, 82: 173–7.

Treherne J. E., Foster W. A. (1980). The effects of group size on predator avoidance in a marine insect. *Animal Behaviour*, 28: 1119–22.

Treichler F. R., van Tilburg D. (1996). Concurrent conditional discrimination tests of transitive inference by macaque monkeys: list linking. *Journal of Experimental Psychology: Animal Behavior Processes*, 22: 105–17.

Trillmich F. (1976). Spatial proximity and mate-specific behaviour in a flock of budgerigars (*Melopsittacus undulatus*; Aves, Psittacidae). *Zeitschrift für Tierpsychologie*, 41: 307–31.

Trivers R. L. (1974). Parent–offspring conflict. *American Zoologist*, 14: 249–64.

Uchino B. N., Caciopoppo J., Kiecolt-Glaser J. (1996). The relationship between social support and physiological processes: a review with emphasis on underlying mechanisms and implications for health. *Physiological Reviews*, 119: 488–531.

Uitdehaag K. A., Rodenburg T. B., van Hierden Y. M., *et al.* (2008). Effects of mixed housing of birds from two genetic lines of laying hens on open field and manual restraint responses. *Behavioural Processes*, 79: 13–18.

van der Velden J., Zheng Y., Patullo B. W., *et al.* (2008). Crayfish recognize the faces of fight opponents. *PLoS One*, 3: 1–7.

van der Wal R. (1998). *Defending the marsh: herbivores in a dynamic coastal ecosystem*. Dissertation, Mathematics and Natural Sciences, University of Groningen, Netherlands.

Van Elzakker M., O'Reilly R. C., Rudy J. W. (2003). Transitivity, flexibility, conjunctive representations, and the hippocampus. I. An empirical analysis. *Hippocampus*, 13: 334–40.

van Oers K., de Jong G., Drent P. J., *et al.* (2004). A genetic analysis of avian personality traits: correlated response to artificial selection. *Behavior Genetics*, 34: 611–19.

van Oers K., de Jong G., van Noordwijk A. J., *et al.* (2005a). Contribution of genetics to the study of animal personalities: a review of case studies. *Behaviour*, 142: 1185–206.

van Oers K., Klunder M., Drent P. J. (2005b). Context dependence of personalities: risk-taking behavior in a social and a nonsocial situation. *Behavioral Ecology*, 16: 716–23.

van Oers K., Buchanan K. L., Thomas T. E., *et al.* (2011). Correlated response to selection of testosterone levels and immunocompetence in lines selected for avian personality. *Animal Behaviour*, 81: 1055–61.

van Schaik C. P., Aureli F. (2000). The natural history of valuable relationships in primates. *In* Aureli F., De Waal F. B. M. (eds.) *Natural Conflict Resolution*. University of California Press, Berkeley, CA, pp. 307–33.

van Schaik C. P., Deaner R. (2003). Life history and evolution in primates. *In* de Waal F. B. M, Tyack P. L. (eds.) *Animal Social Complexity*. Harvard University Press, Cambridge, MA, pp. 5–25.

Vasconcelos M. (2008). Transitive inference in non-human animals: an empirical and theoretical analysis. *Behavioural Processes*, 78: 313–34.

Verbeek M. E. M., Drent P. J., Wiepkema P. R. (1994). Consistent individual differences in early exploratory behaviour of male great tits. *Animal Behaviour*, 48: 1113–21.

Verhulst S., Hut R. A. (1996). Post-fledging care, multiple breeding and the costs of reproduction in the great tit. *Animal Behaviour* 51: 957–66.

Verhulst S., Salomons H. M. (2004). Why fight? Socially dominant jackdaws, *Corvus monedula*, have low fitness. *Animal Behaviour*, 68: 777–83.

Vøllestad L., Quinn T. P. (2003). Trade-off between growth rate and aggression in juvenile coho salmon, *Oncorhynchus kisutch*. *Animal Behaviour*, 66: 561–8.

von Fersen L., Wynne C. D., Delius J. D., *et al.* (1991). Transitive inference formation in pigeons. *Journal of Experimental Psychology: Animal Behavior Processes*, 17: 334–41.

von Holst D. (1986a). Psychosocial stress and its pathophysiological effects in tree shrews (*Tupaia belangeri*). *In* Schmidt T. H., Dembroski T. M., Blümchen G. (eds.) *Biological and Psychological Factors in Cardiovascular Disease*. Springer, Berlin, pp. 508–16.

von Holst D. (1986b). Vegetative and somatic compounds of tree shrews' behavior. *Journal of the Autonomic Nervous System* (Supplement): 657–70.

von Holst D. (1998). The concept of stress and its relevance for animal behavior. *Advances in the Study of Behavior*, 27: 1–131.

Voslamber B., van Winden J., Koffijberg K. (2004). *Atlas van Ganzen, Zwanen en Smienten in Nederland*. SOVON, Beek-Ubbergen, Netherlands.

Voslamber B., van der Jeugd H. P., Koffijberg K. (2007). Aantallen, trends en verspreiding van overzomerende ganzen in Nederland. *Limosa*, 80: 1–17.

Wachtmeister C.-A. (2001). Display in monogamous pairs: a review of empirical data and evolutionary explanations. *Animal Behaviour*, 61: 861–8.

Waldeck P., Andersson M., Kilpi M., *et al.* (2008). Spatial relatedness and brood parasitism in a female-philopatric bird population. *Behavioral Ecology*, 19: 67–73.

Wanker R. (1999). Socialization in spectacled parrotlets (*Forpus conspicillatus*): how juveniles compensate for the lack of siblings. *Acta Ethologica*, 2: 23–8.

Ward A. J. W., Hart P. J. B. (2003). The effect of kin and familiarity in interactions between fish. *Fish and Fisheries*, 4: 348–58.

Ward C., Trisko R., Smuts B. B. (2009). Third-party interventions in dyadic play between littermates of domestic dogs, *Canis lupus familiaris*. *Animal Behaviour*, 78: 1153–60.

Ward S., Bishop C. M., Woakes A. J., *et al.* (2002). Heart rate and the rate of oxygen consumption of flying and walking barnacle geese (*Branta leucopsis*) and bar-headed geese (*Anser indicus*). *Journal of Experimental Biology*, 205: 3347–56.

Warneken F., Tomasello M. (2006). Altruistic helping in human infants and young chimpanzees. *Science*, 311: 1301–3.

Warren S. M., Fox A. D., Walsh A., O'Sullivan P. (1993). Extended parent–offspring relationships in Greenland white-fronted Geese (*Anser albifrons flavirostris*). *Auk*, 110: 145–8.

Wascher C. A. F., Arnold W., Kotrschal K. (2008a). Heart rate modulation by social contexts in greylag geese (*Anser anser*). *Journal of Comparative Psychology*, 122: 100–7.

Wascher C. A. F., Scheiber I. B. R., Kotrschal K. (2008b). Heart rate modulation in bystanding geese watching social and non-social events. *Proceedings of the Royal Society B*, 275: 1653–9.

Wascher C. A. F., Scheiber I. B. R., Weiß B. M., et al. (2009). Heart rate responses to agonistic encounters in greylag geese, Anser anser. Animal Behaviour, 77: 955–61.

Wascher C. A. F., Scheiber I. B. R., Braun A., et al. (2011). Heart rate responses to induced challenge situations in greylag geese (Anser anser). Journal of Comparative. Psychology 125: 116–19.

Wascher C. A. F., Weiß B. M., Arnold W., et al. (2012). Physiological implications of pair-bond status in greylag geese. Biology Letters, 8: 347–50.

Watts D. P., Comenares F., Arnold K. (2000). Redirection, consolation, and policing. In Aureli F., De Waal F. B. M. (eds.) Natural Conflict Resolution. University of California Press, Berkeley, CA, pp. 281–301.

Watts H. E., Holekamp K. E. (2007). Hyena societies. Current Biology, 17: R657–R660.

Weigmann C., Lamprecht J. (1991). Intraspecific nest parasitism in bar-headed geese, Anser indicus. Animal Behaviour, 41: 677–88.

Weiß B. M. (2000). Social support in juvenile greylag geese (Anser anser). Diploma thesis, Institute of Zoology, University of Vienna.

Weiß B. M., Foerster K. (2013). Age and sex affect quantitative genetic parameters for dominance rank and aggression in free-living greylag geese. Journal of Evolutionary Biology, 26: 299–310.

Weiß B. M., Kotrschal K. (2004). Effects of passive social support in juvenile Greylag Geese (Anser anser): a study from fledging to adulthood. Ethology, 110: 429–44.

Weiß B. M., Scheiber I. B. R. (2013). Long-term memory of hierarchical relationships in free-living greylag geese. Animal Cognition, 16: 91–7.

Weiß B. M., Möstl E., Hirschenhauser K. (2005). Within-pair testosterone compatibility as a currency for pairbond quality in greylag geese? Hormones and Behavior, 48: 133.

Weiß B. M., Kotrschal K., Frigerio D., et al. (2008). Birds of a feather stay together: extended family bonds and social support in greylag geese (Anser anser). In Ramirez R. N. (ed.) Family Relations: Issues and Challenges. Nova Science Publishers, New York, pp. 69–88.

Weiß B. M., Kehmeier S., Schloegl C. (2010a). Transitive inference in free-living greylag geese, Anser anser. Animal Behaviour, 79: 1277–83.

Weiß B. M., Kotrschal K., Möstl E., et al. (2010b). Social and life-history correlates of hormonal partner compatibility in greylag geese (Anser anser). Behavioral Ecology, 21: 138–43.

Weiß B. M., Kotrschal K., Foerster K. (2011). A longitudinal study of dominance and aggression in greylag geese (Anser anser). Behavioral Ecology, 22: 616–24.

Whitehead H., Connor R. C. (2005). Alliances. I. How large should alliances be? Animal Behaviour, 69: 117–26.

Whiten A., Byrne R. W. (1988). Tactical deception in primates. Behavioral and Brain Sciences, 11: 233–73.

Whiten A., Byrne R. W. (1997). Machiavellian Intelligence. II. Extensions and Evaluations. Cambridge University Press, Cambridge.

Whitman C. O. (1899). Biological Lectures from the Marine Biological Laboratory of Woods Hole. Ginn & Company, Boston, MA.

Wikelski M., Kays R. W., Kasdin N. J., et al. (2007). Going wild: what a global small-animal tracking system could do for experimental biologists. Journal of Experimental Biology, 210: 181–6.

Wilkinson A., Mandl I., Bugynar T., et al. (2010). Gaze following in the red-footed tortoise (Geochelone carbonaria). Animal Cognition, 13: 765–9.

Wilson A. P., Boelkins C. (1970). Evidence for seasonal variation in aggressive behaviour by *Macaca mulatta*. *Animal Behaviour*, 18: 719–24.

Wilson E. O. (1975a). *Sociobiology: The New Synthesis*. Harvard University Press, Cambridge, MA.

Wilson E. O. (1975b). Communication: origins and evolution. *In* Wilson E. O. (ed.) *Sociobiology*. Harvard University Press, Cambridge, MA, pp. 224–31.

Wingfield J. C., Silverin B. (1986). Effects of corticosterone on territorial behavior of free-living male song sparrows (*Melospiza melodia*). *Hormones and Behavior*, 20: 405–17.

Wingfield J. C., Hegner R. E., Dufty A. M., *et al.* (1990). The 'challenge hypothesis': theoretical implications for patterns of testosterone secretion, mating systems and breeding strategies. *American Naturalist*, 136: 829–46.

With K. A., Balda R. P. (1990). Intersexual variation and factors affecting parental care in western bluebirds: a comparison of nestling and fledgling periods. *Canadian Journal of Zoology*, 68: 733–42.

Witkowski J. (1983). Population studies of the greylag goose *Anser anser* breeding in the Barycz Valley, Poland. *Acta Ornithologica*, 19: 179–216.

Wittenberger J. F. (1981). *Animal Social Behavior*. Duxbury Press, Boston, MA.

Wright R., Giles N. (1988). Breeding success of Canada and greylag geese *Branta canadensis* and *Anser anser* on gravel pits. *Bird Study*, 35: 31–5.

Wyles J. S., Kunkel J. C., Wilson A. C. (1983). Birds, behavior, and anatomical evolution. *Proceedings of the National Academy of Sciences USA*, 80: 4394–7.

Wynne C. D. (1997). Pigeon transitive inference: tests of simple accounts of a complex performance. *Behavioural Processes*, 39: 95–112.

Yamamoto S., Humle T., Tanaka M. (2009). Chimpanzees help each other upon request. *PLoS One*, 4: e7416.

Ydenberg R. C., Prins H. H. T., van Dijk J. (1984). The lunar rhythm in the nocturnal foraging activities of wintering barnacle geese. *Wildfowl*, 35: 93–6.

Yoerg S. I. (1998). Foraging behavior predicts age at independence in juvenile Eurasian dippers (*Cinclus cinclus*). *Behavioral Ecology*, 9: 471–7.

York A. D., Rowell T. E. (1988). Reconciliation following aggression in patas monkeys, *Erythrocebus patas*. *Animal Behaviour*, 36: 502–9.

Zaias J., Breitwisch R. (1989). Intra-pair cooperation, fledgling care, and renesting by northern mockingbirds (*Mimus polyglottos*). *Ethology*, 80: 94–110.

Zann R. A. (1996). *The Zebra Finch: A Synthesis of Field and Laboratory Studies*. Oxford University Press, New York.

Zayan R., Vauclair J. (1998). Categories as paradigms for comparative cognition. *Behavioural Processes*, 42: 87–99.

Zduniak P. (2006). The prey of hooded crow (*Corvus cornix* L.) in wetland: study of damaged egg shells of birds. *Polish Journal of Ecology*, 54: 491–8.

Index

Adelie penguin, 51
adoption, 158, 171, 173
affective arousal, 154
affiliation, 32, 39, 198, 200
affiliative behaviour, 37, 53, 156, 196, 197
African elephant, 105, 177
aggression
    repeated attack, 160, 162, 170
    serial attack, 163,
        see also aggression; repeated attack
    threat, 121, 157
    wing-shoulder fight, 122, 139
aggressiveness, 46, 48–50, 51, 62, 129, 131
agonistic
    encounter, 147, 151–3, 160, 161, 163, 166
    interaction, 32, 111, 116, 122, 129, 157, 160, 163, 166, 170, 181, 199
Alces alces, see European elk
allo-
    feeding, 196, 197
    grooming, 71, 196, 197
    preening, 196, 200
allostatic load/overload, see stress, chronic
Alm river, 12
Almsee, 12, 16
Alpine marmot, 156
American kestrel, 62
analogy, 192, 197
androgens, 50, 77, 80, see also testosterone metabolites
animal model, 134
Anser
    albifrons flavirostris, see white-fronted goose, greater
    anser, see greylag goose

anser domesticus, see domestic goose
caerulescens, see snow goose
canagicus, see emperor goose
cygnoides, see swan goose
erythropus, see white-fronted goose, lesser
fabalis, see bean goose
indicus, see bar-headed goose
rossii, see Ross's goose
Anseranas semipalmata, see magpie goose
Astatotilapia burtoni, see cichlid

bar-headed goose, 7, 75, 94, 97–8, 101, 102, 137
barn swallow, 106
barnacle goose, 23, 73, 94, 100, 101, 106, 107, 122, 134, 136, 137, 164, 166
bean goose, 7
behavioural
    coordination, see pair bond, within-pair coordination
    repertoire, 30, 197
    roles, 67, 87
    synchrony, 67–70, 74, see also pair bond, within-pair coordination
    syndrome, 38, 45,
        see also personality
best of a bad job/situation, 92, 117, 129
bighorn sheep, 47, 154
biparental care, 38, 66, 80, 100
black-capped chickadee, 93
black-legged kittiwake, 74
blue jay, 130
blue-backed manakin, 67
body
    condition, 106, 112
    mass, 150

body (*cont.*)
  posture, 36, 147, 197
  reserves, 168
  size, 122, 163
  temperature, 8, 144
bonding type, 33, 36
brain size, 193–5
brant goose, 6, *see also* brent goose
Branta
  bernicla, see brant or brent goose
  canadensis, see Canada goose
  leucopsis, see barnacle goose
  ruficollis, see red-breasted goose
breeding
  area, 8, 12, 137
  cooperative, 195
  efficiency, 35, 84
  season, 53, 98, 109
  success, 9, 34, 40, 80
brent goose, 137, *see also* brant goose
brood parasite, 99, 101, 102–3
Buldern, 10, 30
butterbur, 186
bystander, 153, 154

Canada goose, 7, 8, 67, 73, 106, 107
Canis lupus familiaris, see dog
Cape Barren goose, 4
carrion crow, 194
Cavia porcellus, see guinea pig
Cebus capucinus, see white-faced
      capuchin
cerebrotype, 194
Cereopsis novaehollandiae, see Cape
      Barren goose
Cervus elaphus, see red deer
Cherax destructor, see crayfish
chimpanzee, 181, 186, 191, 192
Chinese painted quail, 67
Chiroxiphia pareola, see blue-backed
      manakin
Chlorocebus pygerythrus, see vervet
      monkey
cichlid, 181
cognitive adaptations, 193
collateral kin, 105, 114–17
Columba livia, see pigeon
coping style, 45, 54, 60,
      see also personality
Coquerel's dwarf lemur, 117
cortex, 27, 142, 193
corticosterone, 50–2, 58, 84, 167–70,
      see also glucocorticoids;
      hormones
  metabolites, 56, 82, 139, 147, 160
Corvus
  corax, see common raven
  corone, see carrion crow
  corone cornix, see hooded crow
  frugilegus, see rook

monedula, see Eurasian jackdaw
Coturnix coturnix japonica, see Japanese
      quail
crayfish, 179
Crocuta crocuta, see spotted hyena
Cumberland Game Park, 12, 14, 17, 19
Cyanocitta cristata, see blue jay
Cygnus olor, see mute swan

developmental mode
  altricial, 194, 196, 197
  precocial, 8, 101, 173, 187, 194,
      195, 197
divorce, 72, 73, 95
dog, 21, 50, 154, 180
domestic chicken, 62, 132
domestic goose, 4, 5, 28, 29, 80–6,
      87, 195
domestic pig, 47, 49
domestication, 5, 28, 195
dominance hierarchy, 122, 133, 182
dominance rank, 35, 50, 98, 101, 111,
      116, 121, 163, 164, 181
  causes, 122–4, 126–36
  consequences, 136–41
  measurement, 124

Emberiza schoeniclus, see reed bunting
emotions, 154, 191, 192, 197
emperor goose, 62
energy expenditure, 53, 147, 150, 152
enzyme immunoassay, 54, 161
Erbkoordination, see fixed action
      pattern
Erythrura gouldiae, see Gouldian finch
ethogram, 10, 27, 30
ethology, 26, 29
Eulemur mongoz, see mongoose lemur
Eurasian jackdaw, 194, 197
Eurasian oystercatcher, 90
European elk, 122
European rabbit, 50
European starling, 157
evolution, 117, 158, 173, 191–3, 195
evolutionary constraint, 135
Excalfactoria chinensis, see Chinese
      painted quail
extrapair
  copulation, 100
  fertilisation, 93, 99, 100
  partner, 93
  paternity, 100
  young, 93, 101, 103

Falco sparverius, see American kestrel
family
  extended, 97–100, 105, 107, 117,
      163
  multigenerational, 109, 117
  primary, 106, 108, 130, 161

secondary, 106, 107, 108–13, 130,
    163–4, 168
tertiary, 113–14, 163
feeding
    of the flock, 11, 19
    rate, 137, 167
    time, 7, 111, 166
fight–flight response, *see* stress,
    response, acute
fitness
    benefit, 93, 98, 99, 103
    component, 23, 33
    consequence, 52, 137
    inclusive, 103
fixed action pattern, 28
food sharing, 196
Forpus conspicillatus, see spectacled
    parrotlet
freckled duck, 4

Gallus gallus domesticus, see domestic
    chicken
gaze following, 186–7
glucocorticoids, 71, 82, 142, 167,
    see also corticosterone;
    hormones
gosling helper hypothesis, 106, 137
great-crested grebe, 67
great tit, 47, 48, 49, 67
greeting behaviour, 8, 88, 108, 112,
    196, 199
greylag goose
    breeding, 7, 8–9
    distribution, 6
    feeding ecology, 8
    KLF flock, 10–14, 16–19
    KLF research routines, 19–24
    socio-biology, 9–10
    taxonomy, 4
Grünau, 3, 11, 12, 16, 40
guinea pig, 156

habitat fragmentation, 6
Gymnorhinus cyanocephalus, see pinyon
    jay
Haematopus ostralegus, see Eurasian
    oystercatcher
heart rate, 20, 52, 143, 147–54, 157, 180
    beat-to-beat, 144
    beats per minute, 147, 157
    implantation, 143–4
Heinroth, Oskar, 26
heritability, 45, 48, 51, 134–6
herring gull, 154, 157
hippocampus, 195
Hirundo rustica, see barn swallow
Homo sapiens, see humans
homology, 192, 193, 195
hooded crow, 9, 181
hormones

steroid, 21, 54, 55, 57, 61,
    77, *see also* androgens;
    corticosterone;
    glucocorticoids; testosterone
stress, 52, 82, 84, 139,
    see also corticosterone;
    glucocorticoids
human–goose relationship, 4–6
foster parent, 11, 58, 107, 111, 160,
    164–6, 170, 173, 174–5
hand-raising tradition, 14–16
humans, 67
hunting, 5, 24
hypothalamic–pituitary–adrenal axis
    (HPA), 80, 142, 158, 192

imprinting, 10, 28, 174
incubation, 8, 23, 77, 80, 83–4, 101
individual quality, 72, 84, 99
individual recognition,
    see recognition
individuality, 36, 38, 51
instinct, 28, 193
intraspecific brood parasitism (IBP),
    99, 103

Japanese quail, 60, 132

kin recognition, *see* recognition
Konrad Lorenz Research Station
    (KLF), 3, 10, 12, 14, 16, 29

Lagopus lagopus scoticus, see red
    grouse
Larus argentatus, see herring gull
leadership, 75
learning
    ability, 60
    associative, 174, 178
    performance, 58, 59
    task, 60
Lemur catta, see ring-tailed lemur
Lepomis gibbosus, see pumpkinseed
    sunfish
lineal kin, 105, 108–14
locomotion, 58, 147
long-term data, 19, 30, 124, 157,
    200
long-term memory, 185, 195
Lorenz, Konrad, 3, 10, 26–7, 29, 30,
    31, 33, 40, 46, 191
Loxodonta africana, see African
    elephant

Macaca mulatta, see rhesus monkey
magpie goose, 4, 197
Marmota marmota, see alpine marmot
mate choice, 31, 66, 76, 80
matriline, 105, 117, *see also* social
    bonds, female-centred clan

Max Planck Institute for Behavioral
    Physiology, 10, 30
meerkat, 180
Melospiza melodia, see song sparrow
migration, 4, 6–7, 8, 11, 23, 106
Mirza coquereli, see Coquerel's dwarf
    lemur
mixed model, 124, 134, 138, 146
mongoose lemur, 184
mortality, 11, 23–4, 137
motivation
    aggressive, 91, 122, 129
    sexual, 80, 92, 96
motivational
    component, 163, 164
    factor, 60, 152
moult, 8, 9, 19, 77
Mus musculus domesticus, see house
    mouse
mute swan, 122

natural predation, 11

Oberganslbach (OGB), 12–14, 15, 19
Oryctolagus cuniculus, see European
    rabbit
Ovis canadensis, see bighorn sheep
oxygen consumption, 154

pair bond
    duration, 34, 73–4, 79, 129
    formation, 38, 84, 92, 94, 96
    heterosexual, 65, 88, 90, 91, 93, 99,
        100, 129
    homosexual, 90–3, 100, 129
    homosocial, 92, 96, 129
    monogamous, 7, 33–5, 37, 65, 77,
        84, 88, 97, 100, 195
    polygamous, 32, 33, 34, 35, 86,
        93–9, 116, 129
    quality, 67, 76
    secondary partner, 32, 33, 86, 93
    within-pair coordination, 67–70,
        74–6
    within-pair testosterone
        compatibility, 77–80, 84–6
pallium, 27, 193, 194, 195
Pan troglodytes, see chimpanzee
paper wasp, 178
parental care, *see* biparental care
parental effects, 122, 133–6
parent–offspring conflict, 107
partner compatibility
    behavioural, 66–7
    hormonal, 80, 86, *see also* pair
        bond, within-pair testosterone
        compatibility
Parus major, see great tit
personality, 36, 45, 134
    behavioural variation, 46–9

boldness, 48, 49
    correlated behaviours, 49–50
    physiological processes, 52, 53–63
    reactive–proactive, 49, 62
    reproductive success, 53
Petasites hybridus, see butterbur
phylogeny, 4, 192
physical activity, 147, 152, 155
pigeon, 27, 181
pinyon jay, 181
Podiceps cristatus, see great-crested
    grebe
Poecile atricapillus, see black-capped
    chickadee
Polistes fuscatus, see paper wasp
Porphyrio porphyrio, see pukeko
predators, 7, 9, 11, 21, 72, 186
Procyon lotor, see racoon
pukeko, 90
pumpkinseed sunfish, 47
Pygoscelis adeliae, see Adelie penguin

racoon, 105
rat, 27, 50, 181
Rattus norvegicus, see rat
recognition
    individual, 177–80
    kin, 173–6
red-breasted goose, 5
red deer, 105
red fox, 9, 11, 23
red grouse, 132
reed bunting, 100
relatedness, 103, 172
relationship
    dominance, 121, 124, 130,
        see also dominance hierarchy;
        dominance rank
    long-term, 158
    parent–offspring, *see* family
    social, 49, 66, 160, 172, 180, 193
    third-party, 180
    valuable, 193, 196
relationship intelligence hypothesis, 194
reproductive strategy, 99–103
reproductive success, 52–3, 65, 67, 70,
        74, 78, 93, 99, 103, 137, 138–9,
        see also breeding, success
rhesus monkey, 181
ring-tailed lemur, 184
ritualisation, 67
rook, 194
Ross's goose, 7, 100

seasonal
    pattern, 77
    variation, 131, 147
Seewiesen, 10, 30, 31, 33–7
sex ratio, 11, 92, 94, 132, 133
sexual conflict, 38

snow goose, 8, 23, 72, 100, 101, 107, 114, 224
sociability, 48, 49, 50, 52
social
    ally, 72, 92, 127, 151, 152, 156, 157, 167, 170, 197
    behaviour network, 192
    challenge, 71, 86, 150, 156, 193
    cognition, 172, 173, 199
    complexity, 158, 172, 173, 183, 192–3, 196, 199
    environment, 47, 49, 86, 131–3, 147, 157, 185
    intelligence hypothesis, 172
    organisation, 26, 105
    status, 77, 83, 122, 126, 139, 155, 158
social bonds, see also pair bond
    family bond, see family
    female-centred clan, 118, 158
    sibling bond, 116, see also collateral kin
social support, 71, 98, 108, 127, 135, 151, 157, 170, 196
    active, 116, 154, 157, 161–4, 180
    passive, 157, 158, 164–70
    research methods, 161
sociality, 192
song sparrow, 132
spatial proximity, 70, 72, 74, 118, 173
spectacled parrotlet, 110
spotted hyena, 105, 117, 180
Stictonetta naevosa, see freckled duck
stimulus enhancement, 186
stress
    chronic, 143
    HPA axis, see hypothalamic–pituitary–adrenal axis (HPA)
    management, 158, 170, 199
    reduction, 170
    response, 16, 51, 58, 61, 71, 142, 143, 150, 152, 154, 168
        acute, 143
        fast, see stress, sympatho-adrenomedullary system
        slow, see hypothalamic–pituitary–adrenal axis
    social, 142, 151, 156, 157

sympatho-adrenomedullary system (SAS), 142, 157, 192
Sturnus vulgaris, see European starling
subordinate, 10, 52, 134, 137
Suricata suricatta, see meerkat
survival
    adult, 9
    first year, 9, 24, 110, 131, 137
    gosling, 11, 23
Sus scrofa domesticus, see domestic pig
swan goose, 7

Taeniopygia guttata castanotis, see zebra finch
telemetry
    receiver, 144, 146
    transmitter, 143–4
telencephalon, 193, 194
teleost fish, 27
temperament, see coping style; personality
testosterone, 48, 52, 53–63, 77–80, 84–6, see also androgens
    metabolites, 51, 55, 61, 77
time budget, 68
Tinbergen, Nikolaas, 26, 28
trade-off, 62, 121
tradition, 7, 11, 186
transitive inference, 181–5
tree shrew, 156
triumph ceremony, 32, 88, 197, 199
triumph ceremony group (Triumphgemeinschaft), 197
Tupaia belangeri, see tree shrew

vervet monkey, 48
vocalisations, 36, 147, 153, 157
    contact call, 199
    distress call, 175
    vee call, 8, 112
Vulpes vulpes, see red fox

white-faced capuchin, 105
white-fronted goose
    greater, 7, 23, 106, 107, 109, 111, 114, 157
    lesser, 7
wintering area, 9

zebra finch, 61, 84

Printed in the United States
By Bookmasters